HAS THIS EVE
R

"My refrigerator would not cool, so I took my RV to an RV service center. With the refrigerator in operational mode, the tech opened the access door on the outside of the RV and felt to see if the cooling unit was warm at the bottom, which it was, but there was not adequate cooling on the inside of the refrigerator. I was told to throw it away and purchase either a new cooling unit or a complete new refrigerator."

Fortunately, they came to us for a second opinion, and we made the necessary repairs. In many cases, the customer's bill came to less than $100.

This manual is geared to help you troubleshoot an RV refrigerator so that you can do the following:

A. Make minor repairs yourself.
B. Know whether or not your service tech is qualified to make an educated analysis of the problems and is able to do the repairs successfully with the least amount of cost available to you.

Either way, you save money!

OUR RECOMMENDATIONS

Always ask your RV service tech if he/she is RVRN (Recreational Vehicle Refrigeration Network) certified.

If not, give him/her our Web site and encourage them to obtain the education required so you can save money and receive the best warranty in the industry.

AVERAGE JOE'S RV refrigerator troubleshooting AND repair guide

Roger D. & Onna Lee Ford

Library of Congress Control Number: 2009914253

ISBN: Hardcover 978-1-4500-0149-6

Softcover 978-1-4500-0148-9

Ebook 978-1-4500-0150-2

This book was printed in the United States of America.

To order additional copies of this book, contact:

Xlibris Corporation

1-888-795-4274

www.Xlibris.com

Orders@Xlibris.com

70686

REPAIRS CAN SOMETIMES BE THIS SIMPLE

On February 18, 2009, just before the completion of this manual, we received a new refrigerator from an RV dealer. They had condemned it because of a bad cooling unit. During our inspection, we found the heat element had not been properly installed into the heat element sleeve, which caused insufficient cooling.
The problem had nothing to do with the cooling unit.
In approximately sixty seconds, we were able to place the heat element into the heat element sleeve,
and the refrigerator worked great.

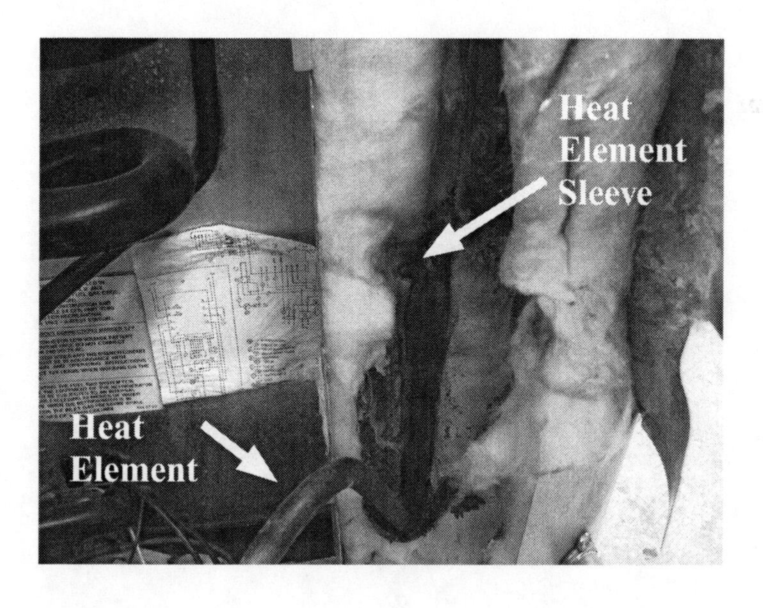

www.rvrefrigeration.com

THE NO. 1 SELF-HELP MANUAL

FOR RV OWNERS AND SERVICE TECHS

VOLUME I
RV REFRIGERATORS

AUTHORS
ROGER D. AND ONNA LEE FORD
Owners and Operators of
Ford RV Refrigeration
Established 1984

The National Association
of RV Parks and Campgrounds says,

The United States twenty-nine million
active campers
and RV owners contribute billions to the
national economy annually.

TABLE OF CONTENTS

TABLE OF CONTENTS

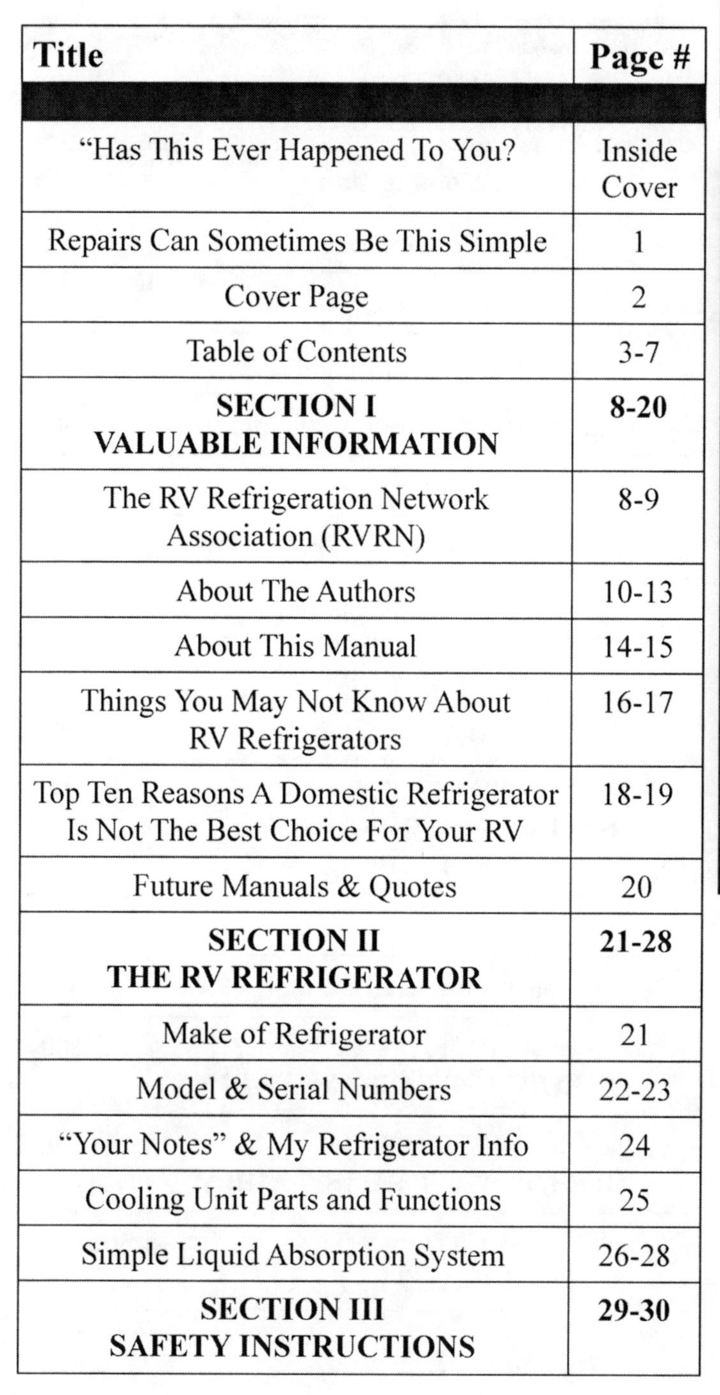

TABLE OF CONTENTS

Thank you. We hope you find this helpful for troubleshooting and minor repairs. If you are looking to enter into the reconditioning industry, there is no experience necessary. We can train you.

SECTION I

VALUABLE INFORMATION

RVRN

THE RV REFRIGERATION NETWORK ASSOCIATION

The RV Refrigeration Network (RVRN) is a new and growing network made up of people who have received certified training in the field of RV refrigerator reconditioning from Ford RV Refrigeration Training Center. These RV refrigerator specialists make themselves progressively available and educated in order to successfully provide a much-needed service to RV owners. In addition, some of them also have other certifications and can assist you with other RV services.

RVRN members come from all walks of life.

Some are HVACR certified, RVDA/RVIA certified, or have owned appliance repair centers. Some were electricians, welders, construction workers, salesmen, and chemical engineers. RV owners have even taken our training because they saw the need for this service in their areas.

> Any RVRN member will service your RV refrigerator and/or recondition your RV refrigerator cooling unit, save you money, and provide you with the RV industry's best warranty.

GO GREEN-GO RNRN

The mission of the RVRN Association is to educate, motivate, and empower the consumers and technicians who have the desire to save money or to provide unsurpassed service and warranties.

VALUABLE INFORMATION

We find working with RVers to be very rewarding and feel they are the most wonderful of all consumers.

RV RESTORATION ENTHUSIASTS

An RVRN refrigerator specialist can recondition your RV refrigerator so that you can keep the original refrigerator in the RV.

To find an RVRN refrigerator specialist nearest you, please check our Web site:

www.rvrefrigeration.com

All RVRN members will have this logo displayed in their establishment.

QUOTE

"Dealers must be entrepreneurs, because one size doesn't fit all when it comes to the RV business."

—*RV Executive Today*, November 2008, 13

I

ABOUT THE AUTHORS

ABOUT THE AUTHORS

By Onna Lee Ford

Since the age of sixteen, Roger D. Ford has worked as a mechanic—often auto mechanic and helping with auto AC repair. In 1978, he and I were married and moved to an apartment above an HVACR business where Roger obtained employment. In the beginning, he was given all the hard-labor jobs. After a few months, he knew he wanted to be a service tech rather than an installer. His boss said if he tried real hard, he might be ready for the service-tech job in three years. However, within a year and a half, his boss realized that if it could be repaired, Roger could make it happen. He gave Roger his own service truck and the title of service technician.

I spent years working as a secretary and taking college courses. Due to health issues, finances, and raising our child, I ended up putting off my education until later years. I did finally reach the goal and have a bachelor's degree in education.

After years of researching absorption refrigeration, Roger decided to go from HVACR service to RV refrigerator reconditioning and invited me to be his partner. In the summer of 1984, we obtained our training and, on October 12, 1984, opened Ford RV Refrigeration.

Within a few months, Roger realized there was a need to make improvements in the procedures we had been taught in order to make the final product better. There wasn't much he didn't try. If the unit didn't work, he wanted to know why, so he took it apart to find the problem. He charged the same units over and over dozens of times trying many things hour after hour, day after day, sometimes staying up days at a time. Though time-consuming, it made him an expert in this field, and thus he created the Ford Procedures of which he then taught to me. Since then we have accomplished the following:

1990 - Began training RV refrigerator repair and cooling unit reconditioning to others throughout the USA and other countries in an attempt to make this service available to all RV owners.

1994 - Wrote "The Ford Procedures," the first step-by-step technical manual on RV refrigerator repair and reconditioning.

2003-2009—Editors from many RV technical magazines engaged us to write articles for their magazines and Web sites.

2005 (November)—Our business, home, vehicles, and all belongings were completely destroyed by a tornado with over 200 mph winds.

2006 (May)—Reopened our doors.
— Awarded Business of the Month by the Chamber of Commerce.
— Became a member of the Recreational Vehicle Dealers Association
— Increased our training to include certification.

2007— Became part of the GO Green movement.
— Developed the RVRN Association. Networked those who have received our certification training.
— Became members of the Better Business Bureau Association.
— Became Kentucky State RV inspectors.

Over the years, our customers have voiced the need for a manual that would enable them to do minor repairs themselves or, at the very least, give them enough education to know whether or not a service tech had been properly educated.

2009— Published the *Average Joe's RV Refrigerator*, which is geared to help both the consumers and service techs.

GO GREEN-Recondition!

Roger D. Ford
President, Instructor, Research and Development, EPA Certified, RVRN Master Certified, RVRN CEO

Onna Lee Ford
Vice President, Instructor, Office Manager, Marketing Director, Web Design, RVRN CEO Assistant

ABOUT THIS MANUAL

Most of the contents found in this manual were taken from a technical service manual, "The Ford Procedures," also written by Roger and Onna Lee and used in correlation with the certified training they offer. This information is based on over thirty years of experience in the refrigeration industry. Twenty-five of those years have been devoted strictly to RV refrigerator reconditioning. It is also from the Ford's thirty years of experience, research and development, as well as answers to questions RV owners have asked over the past twenty-five years.

Pictures in the manual may be of a Dometic or Norcold part. Most of the time, they will be basically the same on either refrigerator.

Every effort has been made to provide you with as much helpful information as possible, easy step-by-step instructions, and other valuable information. Please follow the directions, warnings, and cautions as they are written.

Although every effort has been made to ensure the accuracy and completeness of the information in this book, Ford RV Refrigeration and the authors make no guarantees, stated or implied, nor will they be liable in the event of misinterpretation or human error made by the

reader, or for any typographical errors that may appear.

Work safely with all the tools, wear safety glasses at all times, follow all the warnings, cautions, and notes, and follow all manufacturer's instructions and warnings for all products. Be sure to follow your entire local and state codes if any apply.

QUOTE

"To stay in the black and achieve the maximum earning potential, RV reconditioning needs to be taken seriously. Proper reconditioning of used inventory can create an overall increase in margins."

—*RV Pro* magazine, March 2009

GO GREEN-Recondition!

Things You May Not Know about RV Refrigerators

1. RV refrigerators are environmentally friendly.
    a. The chemicals used in the charge are common compounds found in nature. They are water, ammonia, hydrogen.
    b. The chemicals do not deplete the ozone layer.
    c. They do not contribute to global warming.
    d. They have great thermodynamic appeal, making it more economical.
    e. The distinct odor of the ammonia is a safety factor, which provides early leak detection.
    f. When a leak is detected, the cooling unit can be reconditioned rather than thrown in a landfill and replaced with a new one.
2. The RV refrigerator is user-friendly.
    a. Less maintenance
        i. The refrigerator has few moving parts.
        ii. The refrigeration unit (cooling unit) has no moving parts.
    b. The cooling unit is made of durable steel.
    c. It conveniently operates on LP gas or 110-volt AC or 12-volt DC; therefore, it can operate in transit.
    d. It is designed for safety during transit.
    e. It is insulated to work in outdoor-environment temperatures.

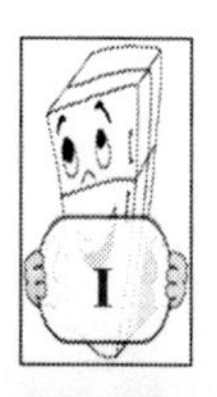

YOU MAY NOT KNOW

3. Why should you have an RVRN refrigerator specialist recondition the cooling unit?
   a. All of no. 1 and no. 2 above.
   b. It is less costly than a new one.
   c. RVRN warranty is better than any warranty in the industry.
   d. Physical-size difference in replacements.
      i. If there is not a replacement cooling unit available for your model of refrigerator, you will have to replace the complete refrigerator. Model sizes are always changing; therefore, you may have to do some remodeling in your RV in order to make it fit properly.

Rumor has it that working with the ammonia and hydrogen is dangerous; however, when the proper procedures are followed, it is no more dangerous than working with common household chemicals or gasoline.

T
O
P

T
E
N

TOP TEN REASONS

Why Domestic Refrigerators (Household) Are Not the Best Choice for Your RV

1. It lowers the resale value of your RV.
2. It has many moving parts leading to more maintenance and repairs.
3. The refrigeration unit is made up of copper and aluminum, which can be damaged when vibrated during transit.
4. The motor in the compressor is also vibration sensitive.
5. It uses more energy.
6. It is insulated to operate in a home with an approximate seventy-degree ambient temperature. If used in an RV, the ambient temps are higher, thus causing a greater use of electricity and wear and tear on the parts of the system.
7. The chemicals used are not environmentally friendly and contribute to global warming.
8. Chemicals used have no obvious odor that helps detect leaks.
9. It operates strictly on 110-volt AC.
10. Physical size will not be the same as the original RV refrigerator, leading to the need to remodel the RV.

GO GREEN-GO RNRN

If you have had an experience with a domestic refrigerator in an RV, feel free to share the information with us.

GO GREEN—Keep refrigerators out of our landfills

TOP TEN

QUOTE

"Environmentally friendly services save the environment and save the consumer money."

—*HVACR Business* magazine, January 2009, 14

QUOTES

Mark, an RVDA/RVIA master certified and RV service center owner in Idaho says, "The Fords are very knowledgeable about new and old refrigerators. If you have a question, they have the answer."

Chauncey, who is an HVAC master certified business owner, says, "This was the best training I've ever attended."

Mike, an appliance repair business owner and an RV owner from Florida, says, "The Fords are great people who look out for the best interest of the customer. You don't find that often today. I highly recommend. Great, great, great!"

RV OWNERS—BE PROUD!

The RV industry leaders are working around the clock to help elevate the service technicians that service your RV.

Ford RV Refrigeration Training Center and the RVRN Association are proud to be part of this team.

Please support our troops
and may God bless you and yours!

SECTION II
THE RV REFRIGERATOR

The make of your refrigerator (see figure 2-1) is located on the front side. Sometimes at the top, sometimes in the middle. On this refrigerator, it's in the middle.

Figure 2-1. Make of refrigerator

Your refrigerator's model and serial numbers will bc located in one of three locations.

1. Inside freezer (bottom or either side) (see figure 2-2)
2. Inside refrigerator (on either side) (see below)
3. On back of refrigerator (see figure 2-3, page 23)

Figure 2-2. Model and serial numbers

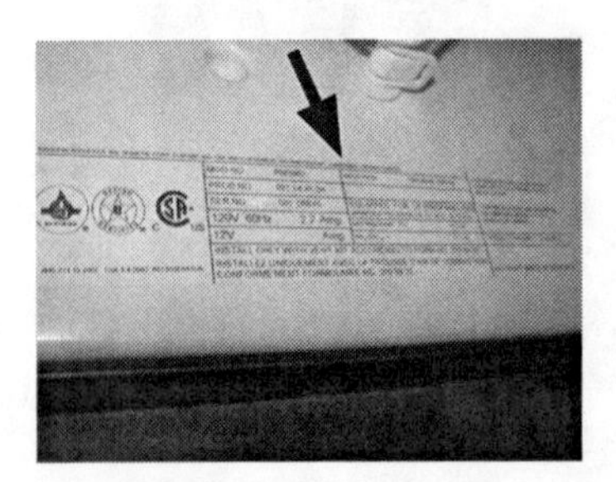

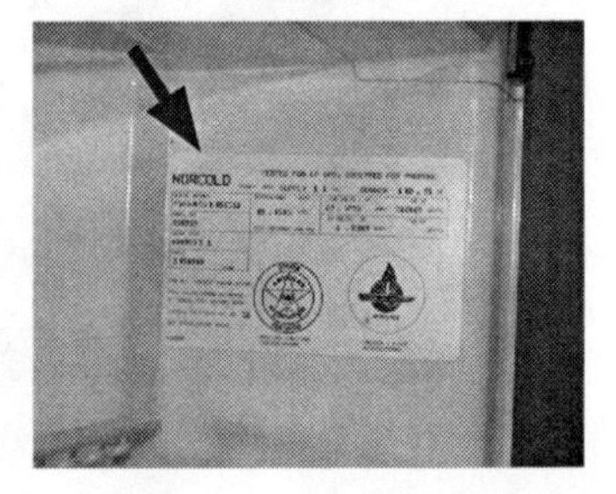

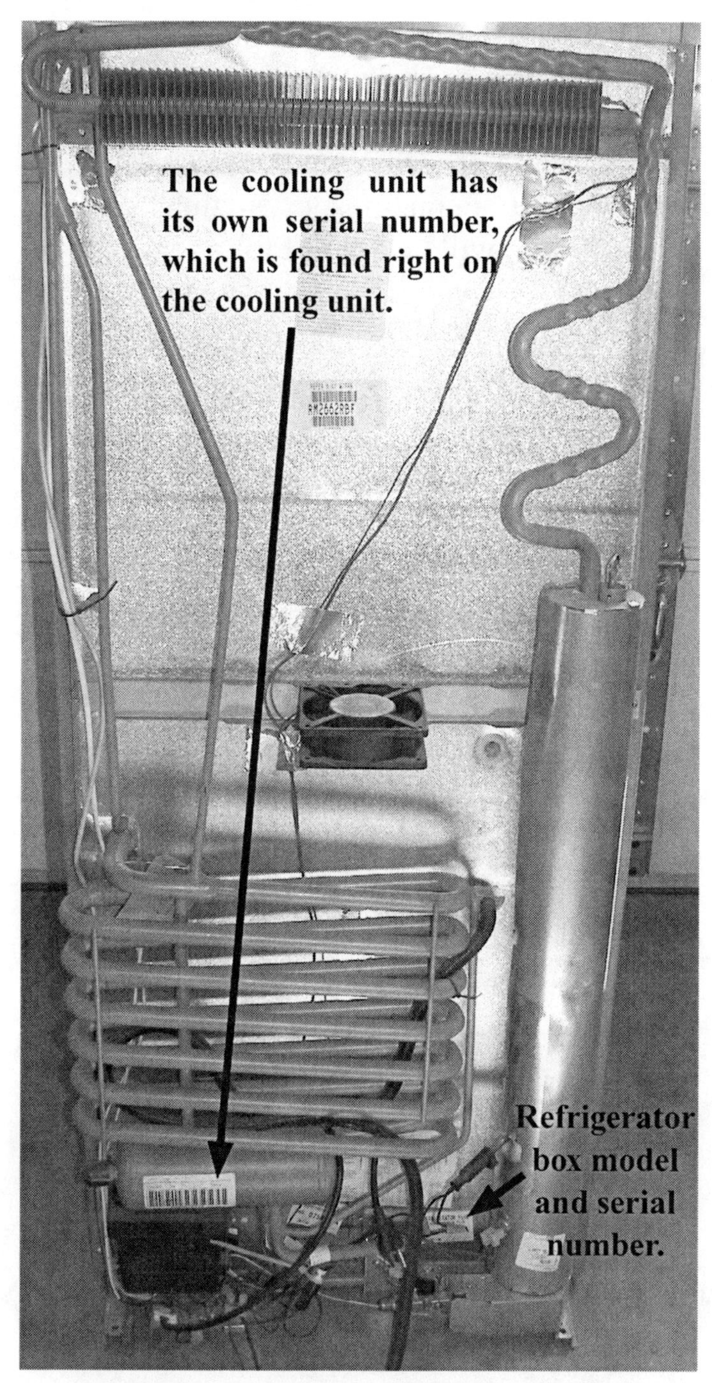

Figure 2-3. Model and serial numbers (continued)

There are places throughout the manual for your notes. This is a good place to put information pertaining to your specific refrigerator, your own how-to ideas, tape in some pictures, or anything else that will be helpful for you.

YOUR NOTES

My Refrigerator Information

Make ____________________

Model ____________________

Serial Number ____________________

Product Number ____________________

QUOTE

"Someone will soon be offering RV refrigerator reconditioning service in your area. Will it be you or the person down the road?"

—Roger D. Ford

COOLING UNIT PARTS AND FUNCTIONS

SIMPLE ABSORPTION SYSTEM

Figure 2-4 on page 26 shows a simple absorption system. Note the absence of a compressor. Here, a heater and boiler are utilized.

This simple absorption system takes its name from the properties of the substances used within the unit—in this case, ammonia and hydrogen. It is the natural tendency of one substance to absorb the other without a chemical change occurring, which allows the system to operate without further compression. Dalton's law states that "the total pressure of a confined mixture of gases is the sum of the pressures of each of the gases in the mixture." In addition, each gas will behave as if it occupied the space alone.

QUOTE

"Myself and others from my area have traveled three hundred miles one way to Ford RV Refrigeration in order to have our refrigerator cooling units reconditioned. We received the best price and the best warranty possible."

—Bill in Illinois, 2008

Figure 2-4. Simple liquid absorption systems

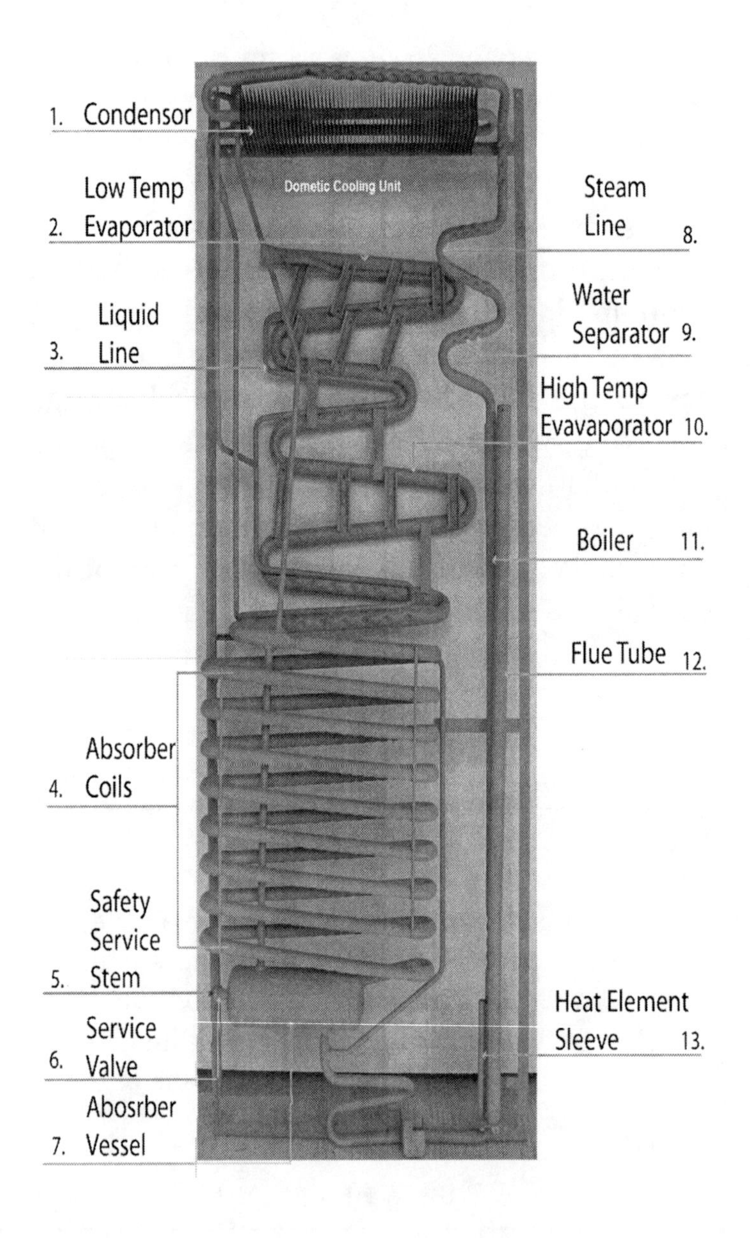

Simple Liquid Absorption System

In the absorption system, ammonia, water, hydrogen, and a rust inhibitor are confined within the unit (see figure 2-4, page 26). Heating the weak solution of ammonia and water at the boiler tube drives it in gaseous form through the boiler tube, carrying the solution to the upper level of the water separator. The liquid settles in the water separator and flows into the absorber vessel. The ammonia vapor rises into the condenser where it will condense into a liquid. The ammonia then flows by gravity into the evaporator. Large quantities of hydrogen gas are fed into the evaporator, allowing the ammonia to evaporate at a low pressure and temperature. Remember, the total pressure of the system is the sum of the vapor pressure of the ammonia and the vapor pressure of the hydrogen. Since the ammonia is condensing, the ammonia vapor pressure drops. It will then naturally try to reach a vapor pressure corresponding to the temperature in the absorber coils and will continue to evaporate until that pressure is achieved. As the ammonia evaporates, it absorbs heat from the food compartment of the refrigerator. The liquid solution, which settled into the separator and flowed into the absorber coil, now meets the mixture of hydrogen gas and ammonia vapor coming from the evaporator. The weak solution absorbs the ammonia vapor while the hydrogen gas—which will not mix with water—is left free. The hydrogen gas, being very light, now rises to the top of the absorber coils and returns to the evaporator. The absorber coils are air-cooled. Cooling the weak solution allows it to reabsorb the ammonia gas from the mixture of ammonia vapor and hydrogen gas. The heat liberated during this stage is removed by the air-cooled coils in order for refrigeration to continue. The ammonia liquid and water mixture then return to the boiler tube where the cycle begins again.

GO GREEN-Keep refrigerators out
of our landfills!

SECTION III

SAFETY INSTRUCTIONS AND EMERGENCY PROCEDURES

PRECAUTIONS

There are specific safety precautions contained in this manual that must be strictly followed by all involved. These precautions precede hazardous operations as *warnings* and *cautions* as follows:

Warning—refers to an operational procedure or step, which, if not strictly adhered to, could result in bodily injury.

Caution—refers to an operational procedure or step, which, if not strictly adhered to, could result in equipment damage.

Note: You should always wear eye protection and have eyewash nearby, especially when working with the cooling unit. Though an accident has never happened to us or, to our knowledge, to anyone we have trained, one can happen, especially when proper procedures are not followed. If the ammonia gets in your eyes, you have approximately fifteen seconds to completely flush them or permanent damage may occur. If an accident does occur, consult your physician right away.

Anyone making any alterations or modifications to their RV or refrigerator should make sure the methods used are in line with the federal, state, and local codes.

SAFETY

NOTES

QUOTE

"I would like to thank Ford RV for the opportunity to become one of their recommended reconditioners for the RVRN. This was an education I will never forget and the best investment I ever made."

—Barry, campground owner in Arkansas, 2008

SECTION IV
TROUBLESHOOTING

The first step in troubleshooting any refrigerator is to determine whether or not the problem lies within the cooling unit or a control.

TROUBLESHOOTING THE COOLING UNIT

1. Visual Inspection

Inspect unit for any yellow coloring. Yellowing is caused by a rust inhibitor and is usually found in the boiler pack area (see figure 4-1). (Although yellowing can occur elsewhere, this is the most visible area with the cooling unit.)

Figure 4-1. Yellowing in boiler area

TROUBLESHOOTING

2. Nasal Inspection

WARNING—Extinguish ALL
OPEN FLAMES!

Hydrogen mixed with saturated ammonia vapor may be escaping from the cooling unit and may be flammable.

Open refrigerator doors and check for ammonia odor. If odor is present, this indicates a leak in the evaporator section. Leave the doors open on the refrigerator and open windows in the RV to safely dissipate the hydrogen and odor.

If either of these inspections indicate a leak in the cooling unit, contact an RVRN refrigerator specialist to have your cooling unit reconditioned.

Even if a leak is not notable with these two inspections, one might still be present. No yellowing around the boiler indicates the leak is above the liquid line. No ammonia odor indicates that odor has dissipated from where the leak occurred, and time has allowed the hydrogen and ammonia odor to escape.

COOLING UNIT OR CONTROL

If the visual or nasal inspections do not indicate a problem with the cooling unit, you need to bypass the controls. This will tell you if the problem is in fact with the cooling unit or if there is a control problem.

3. Probe Inspection

   a. Bypass controls by connecting the heat element directly to 110-volt AC to ensure the heat element is operating properly (see figure 4-2). This allows for the testing of the cooling unit without interference from other controls.

NOTE NO. 1

Do not troubleshoot cooling unit using 12-volt connection or LP gas. A 110-volt AC ensures accurate results.

Figure 4-2. Bypassing the controls

b. To ensure that the heat element is functioning properly and is the required wattage for the unit being serviced, do the following:

   i. Use a multimeter to check the voltage going to the refrigerator from the receptacle.
   ii. Again, using the multimeter, put the amp probe around one wire of the heat element. As an example, we will say the amp reading is 2.6 amps (see figure 4-3).

Figure 4-3. Amp probe around one heat element wire.

iii. Using Ohm's law, take the voltage and multiply the 2.6 amps.

Example:

volts × amps = watts

110 volts × 2.6 amps = 286 watts

iv. You are allowed a 10 percent fluctuation in either direction on the wattage. So if the heat element is 300 watts, the 286 watts is within the 10 percent and the heat element is good. If the wattage is out of the 10 percent, refer to pages 116-191 to find your refrigerator model's correct heat element wattage.

Now that we know we have a heat element that is functioning properly, we can move to the next troubleshooting step.

c. If within two minutes of bypassing the controls, a rapid boil is noted coming from the boiler, the unit has a leak and has lost the gas portion (pressure) of the charge. Contact an RVRN refrigerator specialist to have your cooling unit reconditioned.

d. If no rapid boil is noted, level the refrigerator in the freezer compartment (see figure 4-4) and allow unit to run for approximately one hour. It is important that when the unit is operating, it is level side to side and front to back.

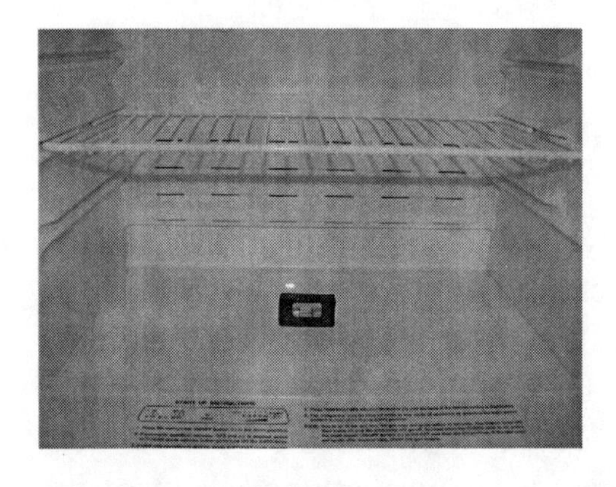

Side to side

Figures 4-4. Leveling the refrigerator

Front to back

OTHER PROBLEM AREAS

Circumstances that would cause the refrigerator unit to be 32°F or above when the controls are bypassed:

a. Unit is unlevel (see figure 4-4, page 36).

b. Poor ventilation (see figure 4-5) behind the refrigerator in the RV. Birds, mud daubers, squirrels, and other critters like to make nests around the vent behind your refrigerator.

    i. Open the refrigerator access door on the outside of the RV. With your head in as far as possible, look up between the back of the refrigerator and the RV wall. For proper ventilation, you need to be able to see daylight.

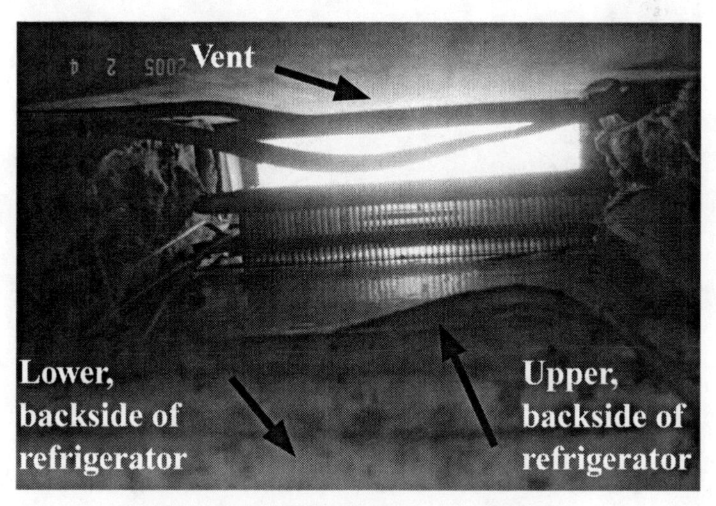

Figure 4-5. Proper ventilation

c. Condensation drain tube is open and needs to be closed (see figures 4-6-A and B).

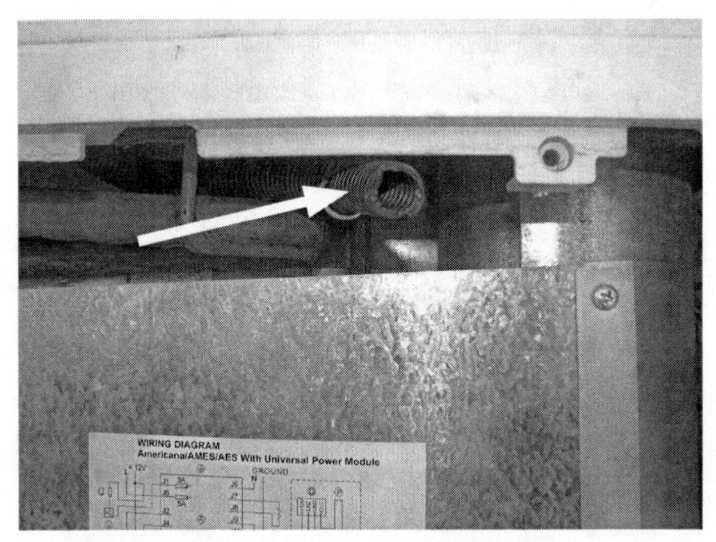

Figure 4-6-a. Open condensation drain

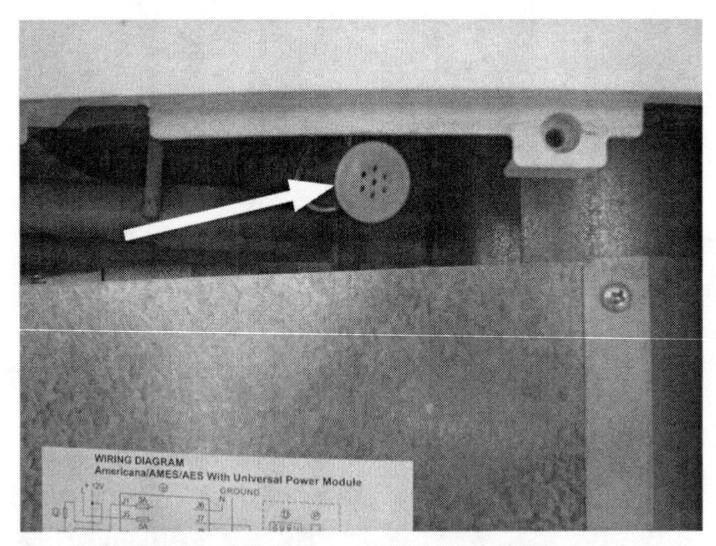

Figure 4-6-b. Closed condensation drain

d. There is insufficient insulation between the boiler and the cabinet (see figure 7-2, page 101).

e. High pressure leak—characterized by holes that are so small they do not completely release all pressure from the cooling unit. The cooling unit needs to be reconditioned. Contact an RVRN refrigerator specialist.

f. Heat element is in incorrect wattage or no longer functions (refer back to pages 34-35).

g. Door gasket needs to be replaced (see figure 4-8, page 41).

4. Temperature

After one hour, some warming should be noted in the bottom absorber coil, cooling observed in the freezer compartment, and considerable warming present at the top of steam line, just before condenser. If these conditions are present, place a thermometer in refrigerator compartment (see figure 4-7, page 40) and allow to run overnight. Check refrigerator temperature the next day. Regardless of ambient temperature, thermometer in refrigerator should read below 32°F.

Figure 4-7. Thermometer in refrigerator compartment

NOTE NO. 2

In the heat of summer, when ambient temperature is 90°F or greater, the refrigerator temperature may only read in the high 20s. In cooler seasons, when temperature drops significantly, that same refrigerator's temperature may read significantly cooler. Ambient temperature plays a big part in these types of cooling units.

QUOTE

"I have serviced more refrigerators in one month at Ford RV Refrigeration than I did in an entire year at the RV Super Center."

—Ford RV Refrigeration former mobile service manager

5. Door Gasket

Check freezer and refrigerator compartment door gasket for signs of wear (see figure 4-8). Often the lower part of the door gasket is the trouble spot. If visually the gasket looks good, check the seal by using a dollar bill. This is done by opening the refrigerator door and inserting a dollar bill between the gasket and the cabinet. Close the door and pull the dollar away from the gasket. There should be some resistance showing a good seal. If there is little or no resistance, this indicates a poor door seal in this area. Repeat this step around the entire door gasket on both doors. Replace the door gasket(s) if necessary. If the door gasket seal is not suspect and, after running unit overnight, the temperature inside is still above 32°F, a high pressure leak may be present.

Figure 4-8. Checking the door gasket

The next few pages cover general troubleshooting. If at any point you need to remove the refrigerator from the RV, refer to the steps starting on page 91.

YOUR NOTES

QUOTE

"Mr. Ford demonstrates extensive knowledge, bearing proof of a career dedicated to the advancement of gas refrigeration."

—Carlos, a chemical engineer and instructor from Texas, 2008

GENERAL TROUBLESHOOTING

SYMPTOMS	POSSIBLE PROBLEM AREAS *Pictures on pages 50-68*	CAUSES/REMEDIES (** means take to an RVRN refrigerator specialist)
Lights on upper board are not on	Fuse (DC) Wiring DC voltage Upper circuit board Lower circuit board	Blown/incorrect—needs to be replaced Loose/wired incorrectly (see p. 51) Voltage is too low or too high ** **
Lights on upper board are on	Thermistor Fuse (AC) Wiring DC voltage Upper circuit board Lower circuit board	Loose / wired incorrectly Blown/incorrect—needs to be replaced Loose / wired incorrectly Diagram voltage is too low or too high ** **

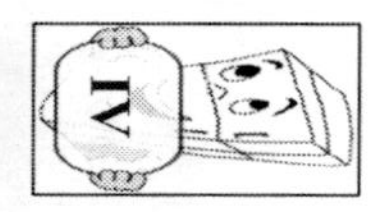

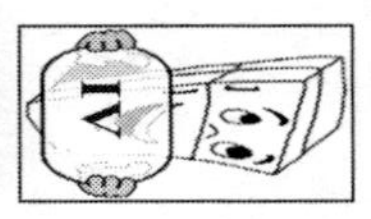

GENERAL TROUBLESHOOTING

SYMPTOMS	POSSIBLE PROBLEM AREAS *Pictures on pages 50-68*	CAUSES/REMEDIES (** means take to an RVRN refrigerator specialist)
Operates on all modes except on AC	AC voltage Heat element Fuse Wiring Upper circuit board Lower circuit board	Voltage is too low or too high Wrong wattage or has gone bad—needs to be replaced or see p. 1 Blown/incorrect—needs to be replaced Loose / wired incorrectly ** **
Operates on all modes except on DC	DC voltage Heat element Fuse Wiring Upper circuit board Lower circuit board	Voltage is too low or too high Wrong wattage or has gone bad—needs to be replaced Blown/incorrect—needs to be replaced Loose / wired incorrectly ** **

GENERAL TROUBLESHOOTING

SYMPTOMS	POSSIBLE PROBLEM AREAS *Pictures on pages 50-68*	CAUSES/REMEDIES (** means take to an RVRN refrigerator specialist)
Operates on all modes (continued) except gas	LP gas	Turned off / disconnected / empty
	Igniter	Needs to be replaced or cleaned
	Electrode	Needs to be cleaned/adjusted/replaced
	Gas valve	Needs to be replaced
	Solenoid	Needs to be replaced
	Voltage cable (110 volt)	Needs to be replaced
	Wiring	Loose / wired incorrectly
	Upper circuit board	**
	Lower circuit board	**
Unsatisfactory temperatures in all modes	Poor ventilation	Area behind refrigerator—lower vent to upper vent
	Thermistor	Loose / wired incorrectly
	Leveling	Refrigerator is unlevel—place level in freezer
	Door gasket	Not getting a good seal—needs to be replaced
	Cooling unit	Has lost the charge—****needs to be reconditioned

GENERAL TROUBLESHOOTING

SYMPTOMS	POSSIBLE PROBLEM AREAS *Pictures on pages 50-68*	CAUSES/REMEDIES (** means take to an RVRN refrigerator specialist)
Satisfactory Temperatures in All Modes		
Except AC	AC voltage Heat element Lower circuit board	Voltage is too low or too high Wrong wattage or has gone bad—needs to be replaced **
Except DC	DC voltage Heat element Wiring Lower circuit board	Voltage is too low or too high Wrong wattage or has gone bad—needs to be replaced Loose / wired incorrectly **
Except gas	LP gas Burner Orifice Flue cap Flue baffle Flue tube Lower circuit board	Turned off / disconnected / empty Needs to be cleaned / needs to be replaced Needs to be cleaned / needs to be replaced Needs to be adjusted / needs to be replaced Needs to be adjusted / needs to be replaced Needs to be cleaned (should be done annually) **

GENERAL TROUBLESHOOTING

SYMPTOMS	POSSIBLE PROBLEM AREAS *Pictures on pages 50-68*	CAUSES/REMEDIES (** means take to an RVRN refrigerator specialist)
Freezes	Thermistor Upper circuit board Lower circuit board	Loose / wired incorrectly ** **
The preset mode changes	DC voltage Wiring Upper circuit board Lower circuit board	Voltage is too low or too high Loose / wired incorrectly ** **
The check light stays on	LP gas Gas valve Burner Orifice Solenoid Thermocouple Wiring Lower circuit board	Turned off / disconnected / empty Needs to be replaced Needs to be cleaned / needs to be replaced Needs to be cleaned / needs to be replaced Needs to be replaced Needs to be cleaned / adjusted / replaced Loose / wired incorrectly **

GENERAL TROUBLESHOOTING

SYMPTOMS	POSSIBLE PROBLEM AREAS *Pictures on pages 50-68*	CAUSES/REMEDIES (** means take to an RVRN refrigerator specialist)
The icemaker won't start up	AC voltage Icemaker cycle Water valve Arm	Voltage is too low or too high Replace icemaker Needs to be replaced Is in the up position
The icemaker does not make ice	AC voltage Water Valve Icemaker Cycle	Voltage is too low or too high Needs to be replaced Replace icemaker
Icemaker won't stop making ice	Shut off Arm	Needs to be adjusted/replaced
Icemaker makes ice, but not enough	Cube size Icemaker cycle Mold thermostat	Replace water valve Replace icemaker Replace icemaker

GENERAL TROUBLESHOOTING

SYMPTOMS	POSSIBLE PROBLEM AREAS *Pictures on pages 50-68*	CAUSES/REMEDIES (** means take to an RVRN refrigerator specialist)
Water is coming out more than normal	Water valve Water fill adjustment	Replace Needs to be adjusted
Ice cubes have frozen onto electric blade	Water valve Water fill adjustment	Needs to be adjusted Needs to be adjusted
Icemaker is hooked up properly but no water comes out	Water Valve Water	Check voltage Check flow

QUOTE

"The ability to recognize trends and jump on them early is what separates good businesses from the super successful."

—*RV Trade Digest* magazine, January-February 2007, 8

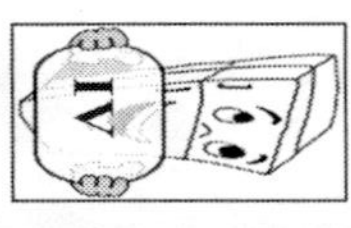

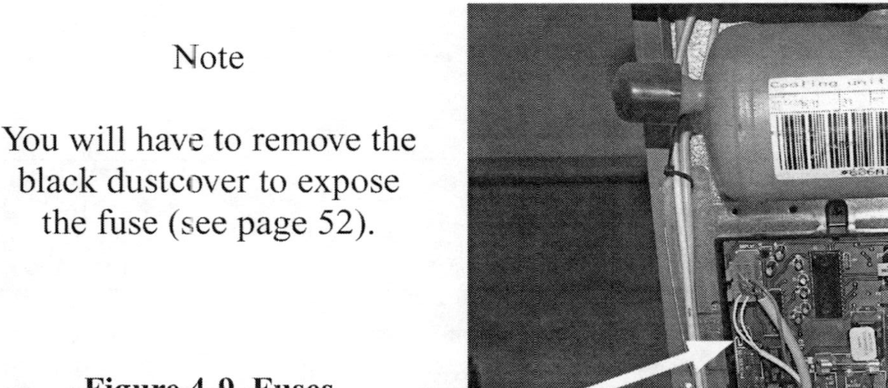

Note

You will have to remove the black dustcover to expose the fuse (see page 52).

Figure 4-9. Fuses (in lower board)

Most lower boards have a 5-amp AC fuse and a 3-amp DC fuse.

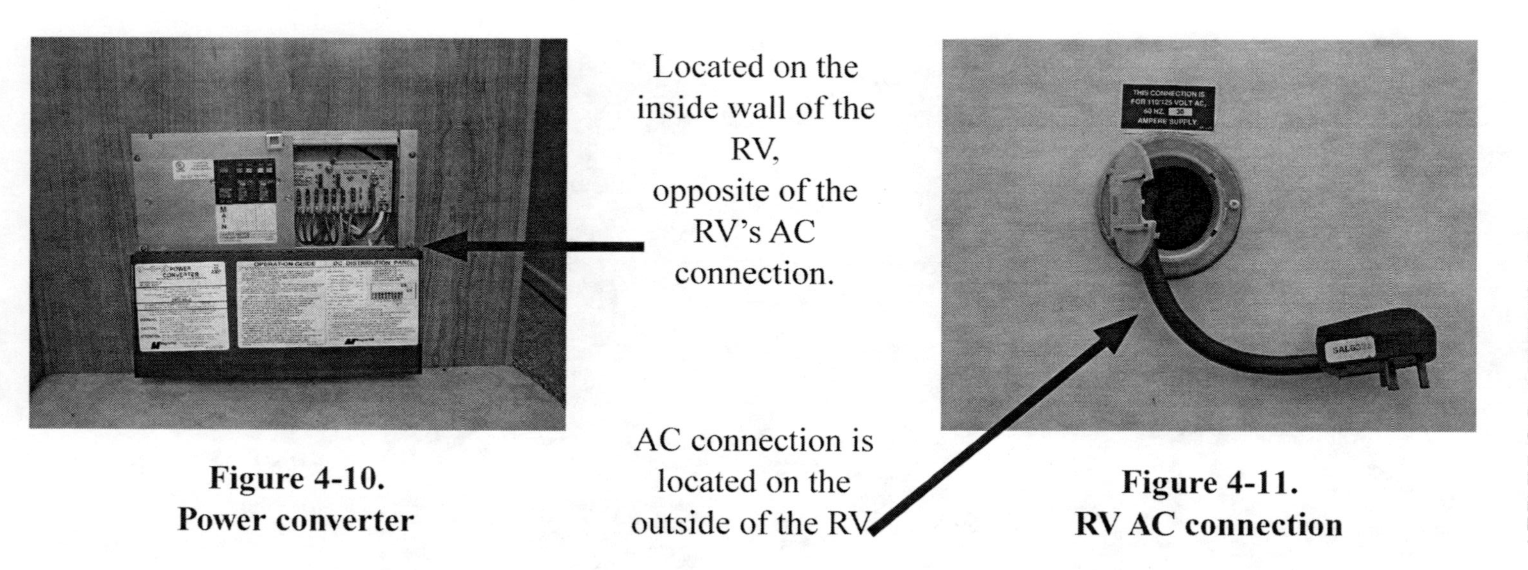

Figure 4-10.
Power converter

Figure 4-11.
RV AC connection

If the RV is plugged into 110 volts, the refrigerator is in the AC mode, and the fuses on the refrigerator are functioning properly, check the 12-volt fuse in the converter.

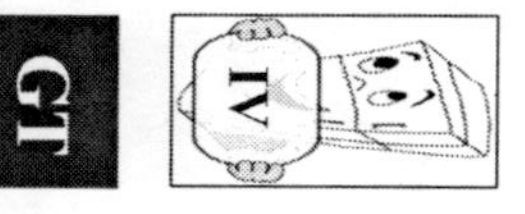

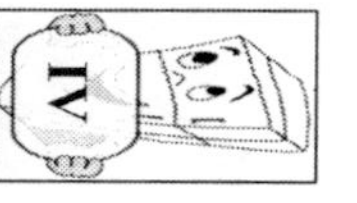

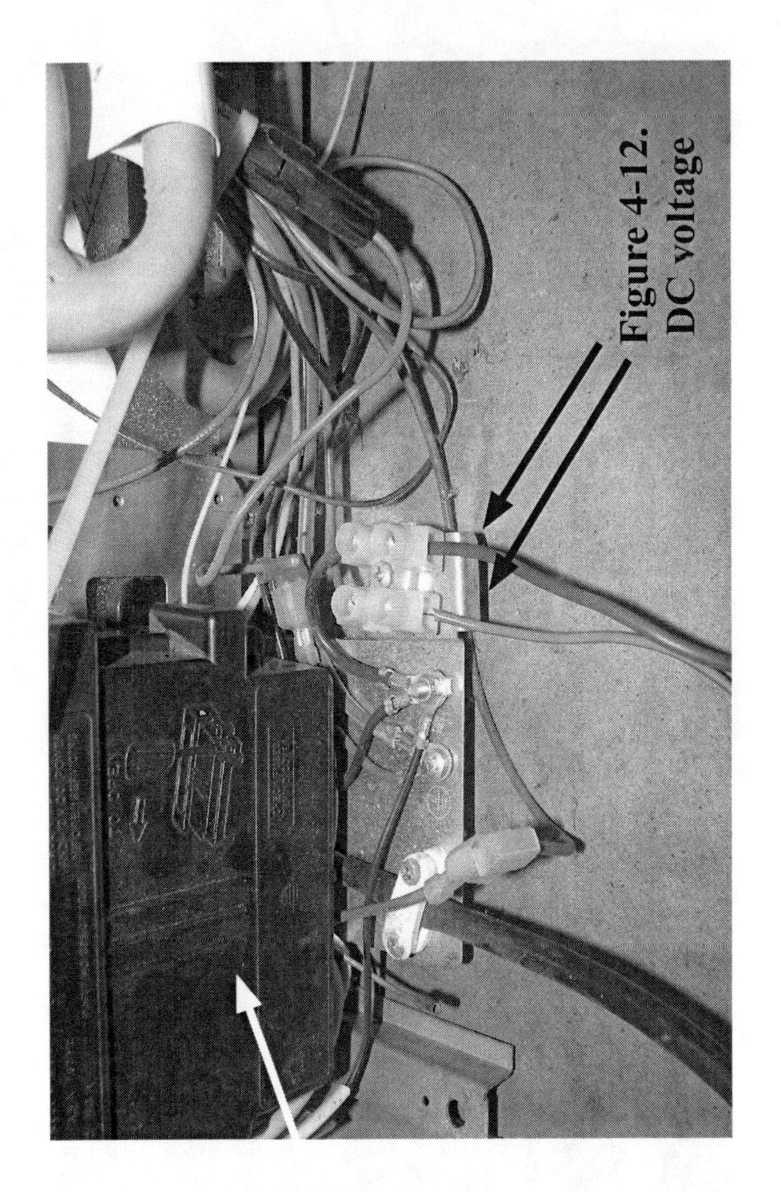

Figure 4-12. DC voltage

Black dustcover

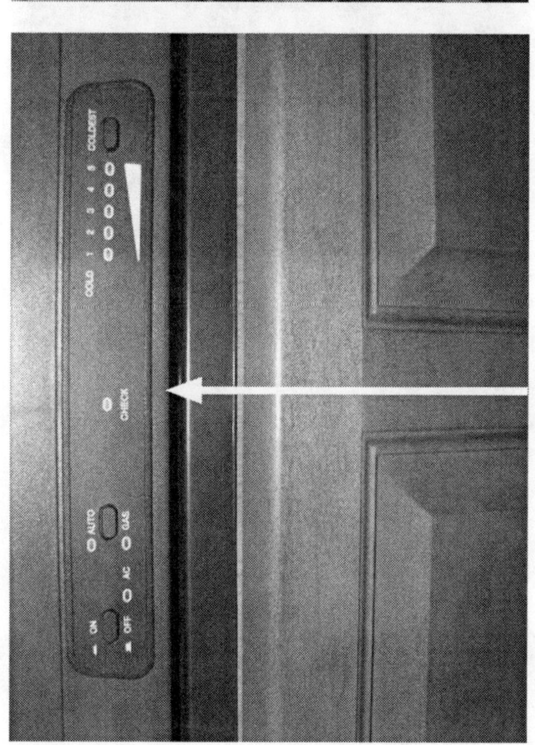

Figure 4-13. Upper Ciruit Board

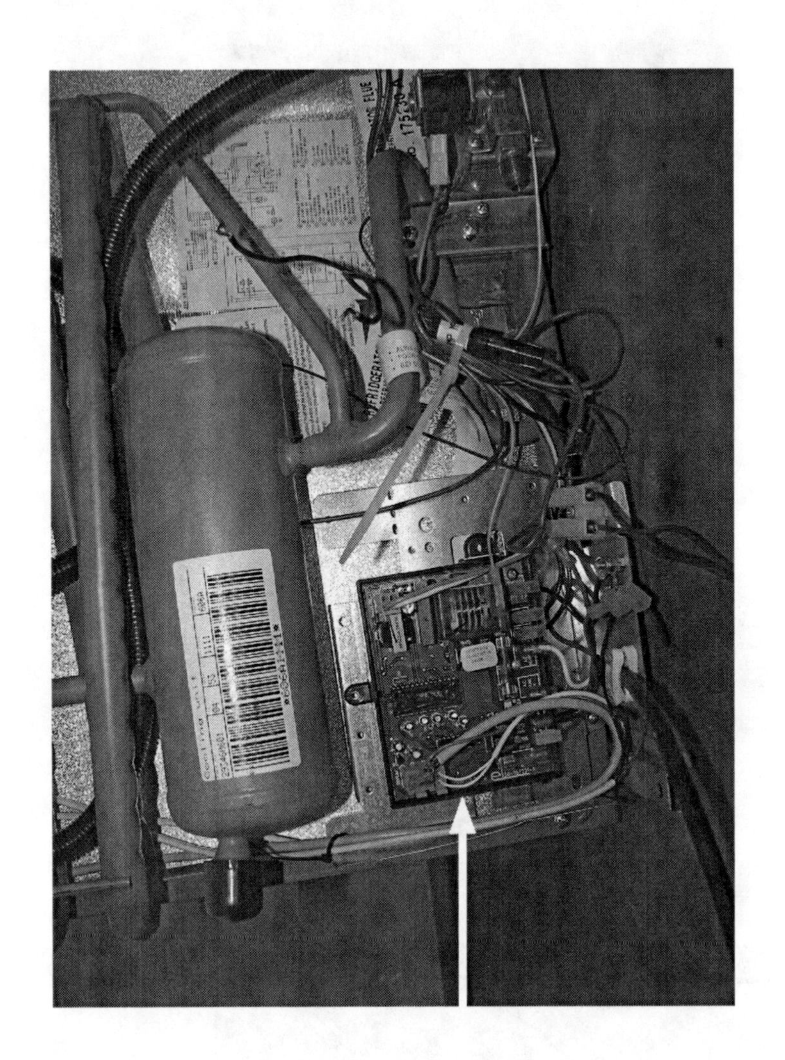

Figure 4-14. Lower circuit board, Dometic

Figure 4-15. Thermistor (temperature sensor)

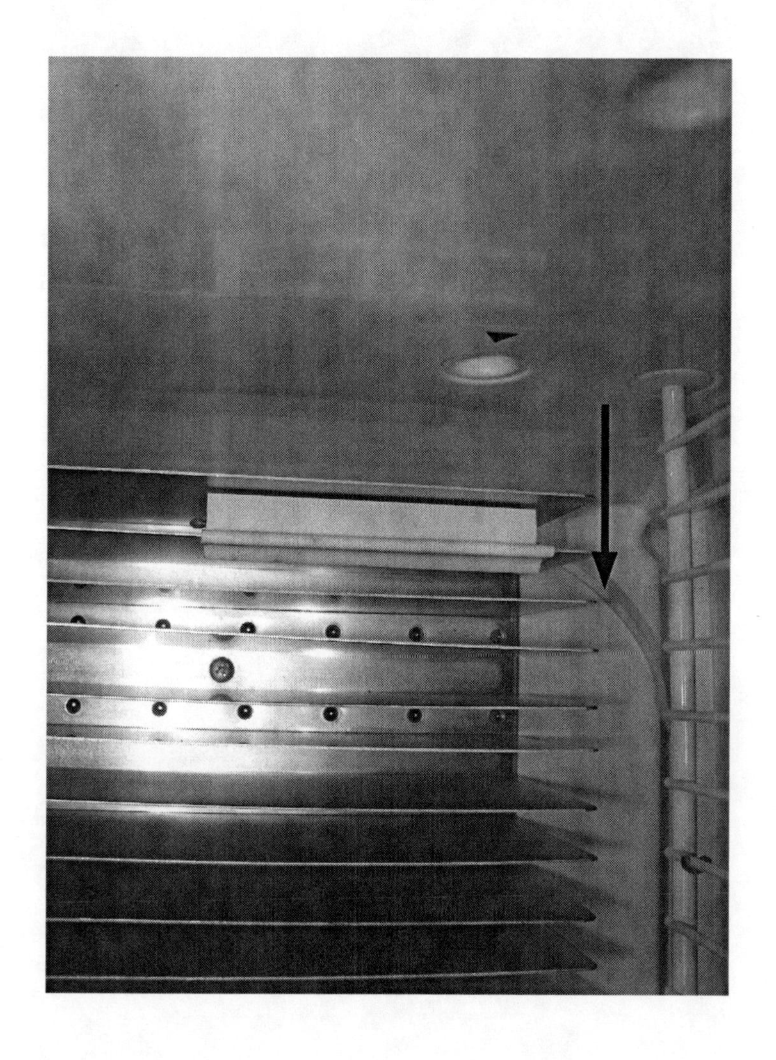

Figure 4-16.
AC voltage (refrigerator)

(Found behind refrigerator at outside access door)

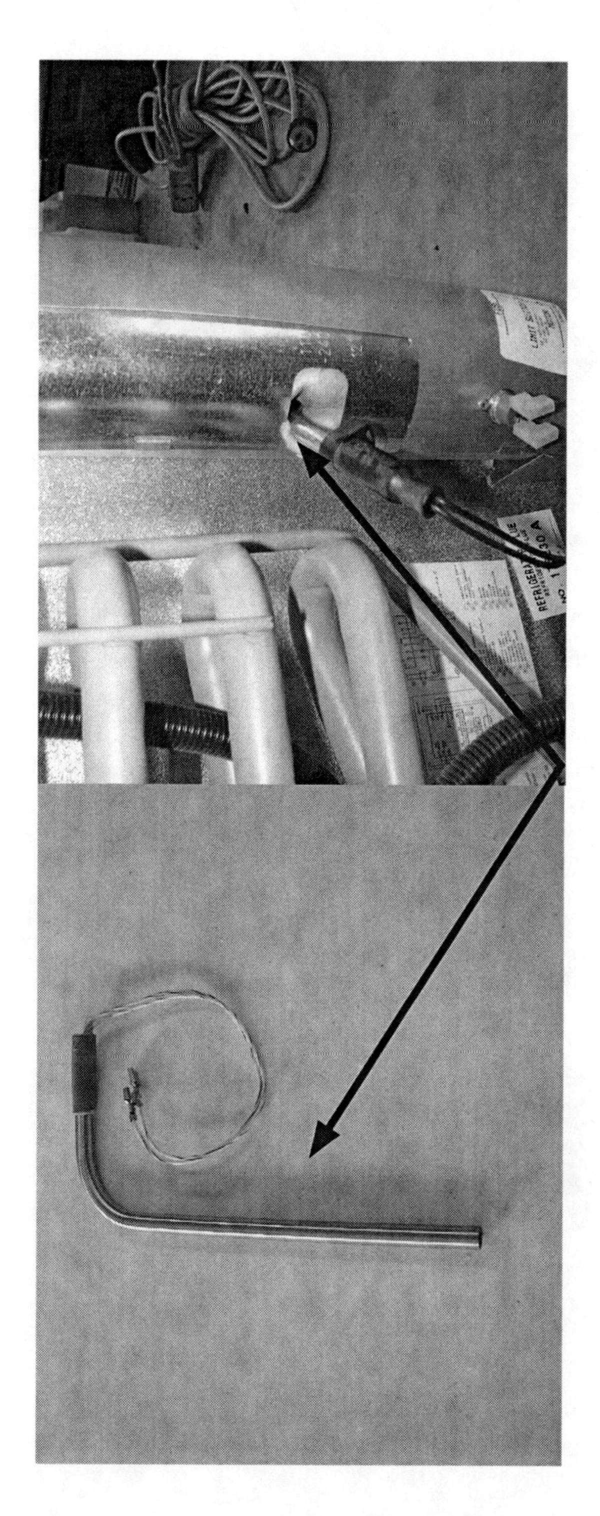

Figure 4-17. Heat element

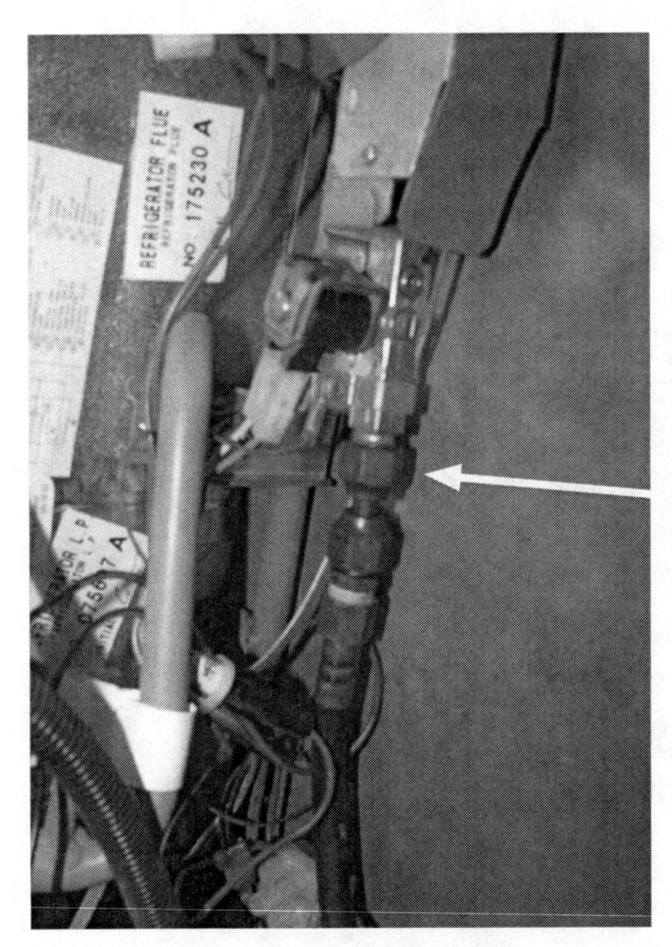

Figure 4-18. LP gas connection

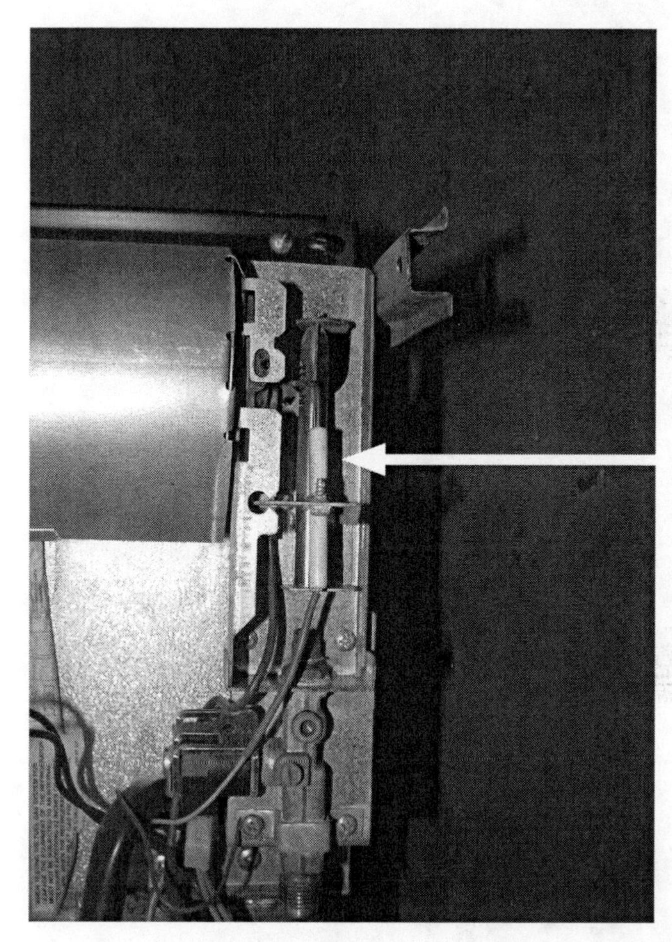

Figure 4-19. Electrode igniter

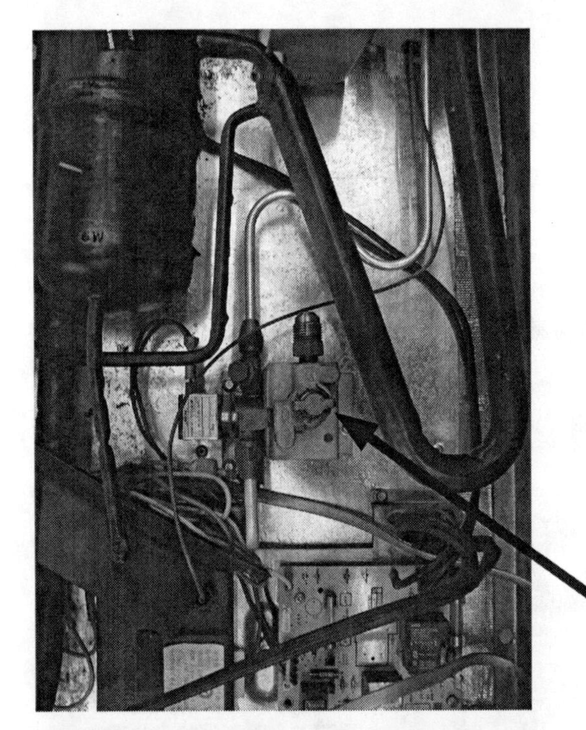

Figure 4-20. Gas valve

Figure 4-21.
Gas solenoid

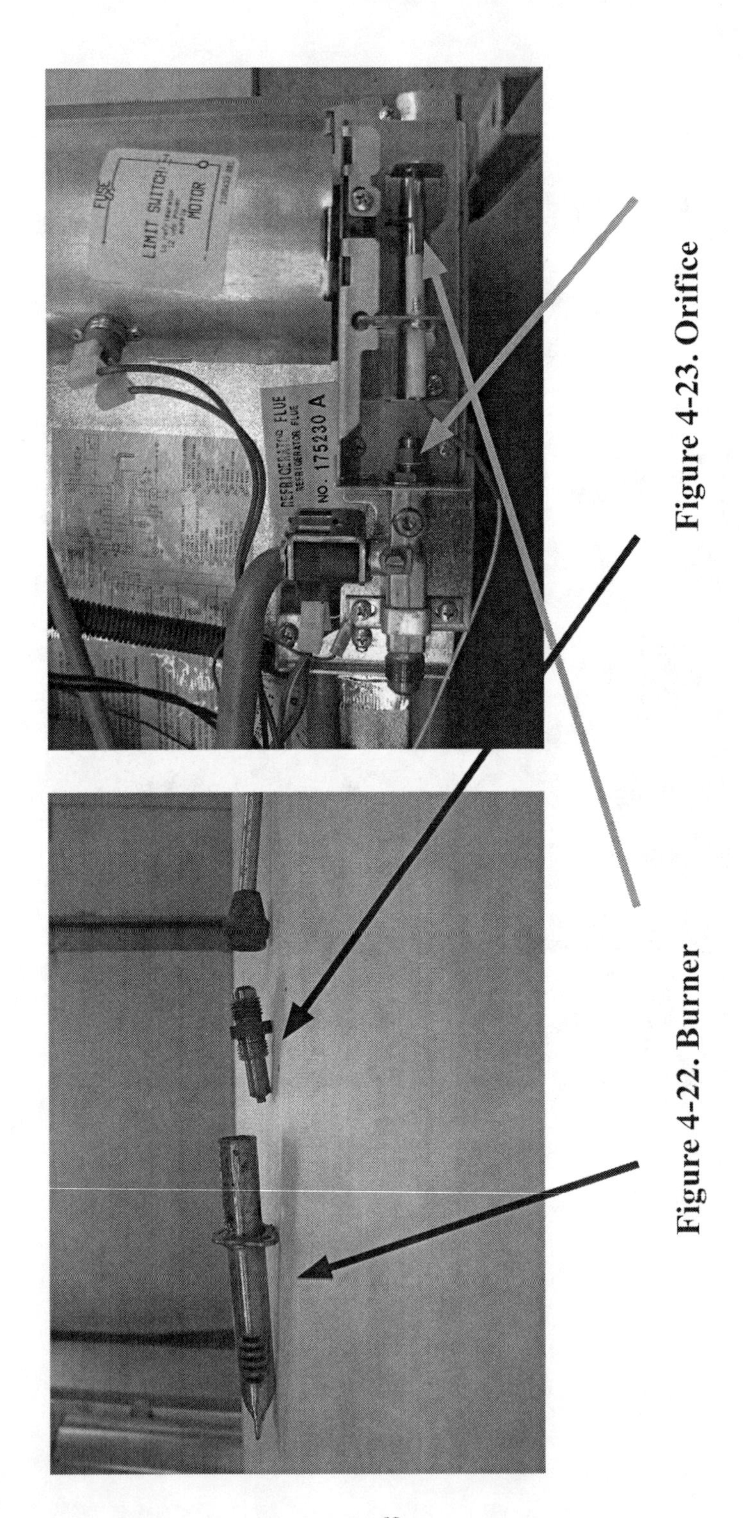

Figure 4-22. Burner

Figure 4-23. Orifice

Figure 4-24. Flue cap

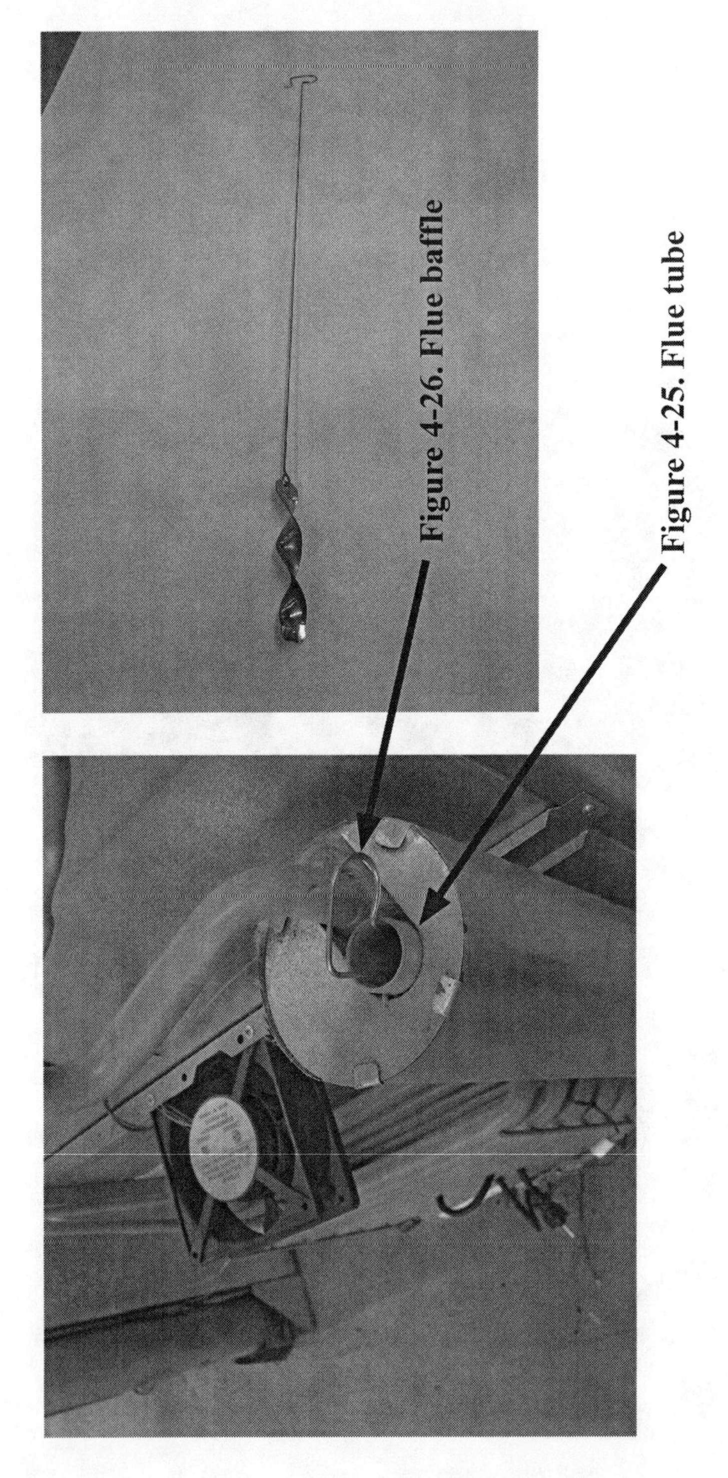

Figure 4-26. Flue baffle

Figure 4-25. Flue tube

Located above
the burner

Figure 4-27. Thermocouple

Figure 4-28.
Humidity switch

Recommend operating in normal operation.

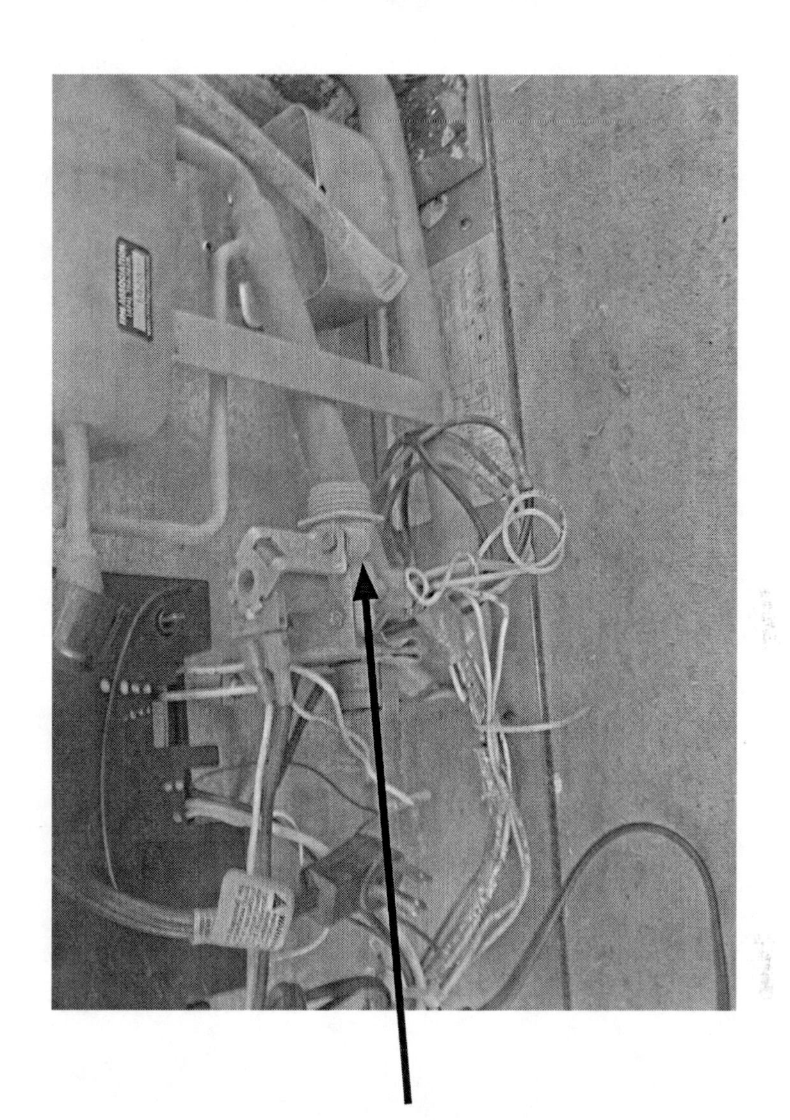

Figure 4-29. Water valve for icemaker

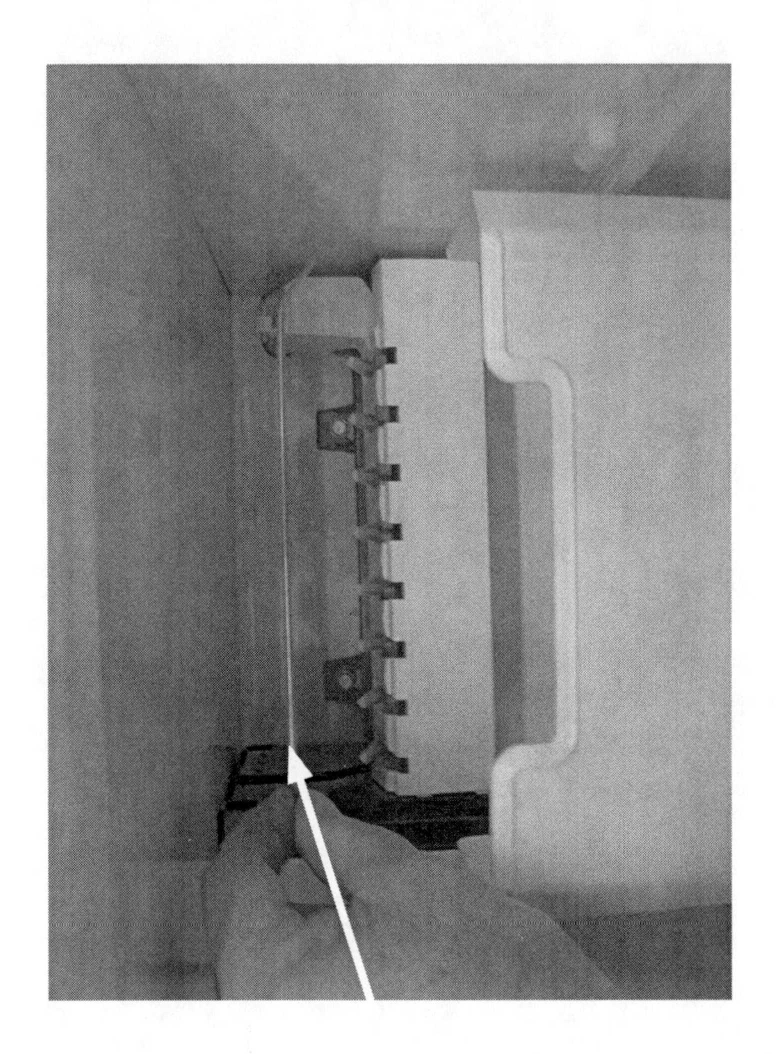

Figure 4-30.
Icemaker arm

GET EDUCATED

"Learn more about the
RV refrigerator
reconditioning
industry."

Go green!
Help save our
environment.

Join the
RV refrigeration
network association.

RV owners, you now have options!

You don't have to replace that expensive refrigerator or its cooling unit in your RV. Have yours repaired by a certified RVRN refrigerator specialist!

Benefits

Saves you money
Receive the industry's best warranty
Increases the resale value of your RV
Average electrical - 2 amps usage

Contact us to find the RVRN service center nearest you.
www.rvrefrigeration.com

GO GREEN!
Chemicals in these units are friendly to the environment.

Keep refrigerators and cooling units out of our landfills.

WARRANTY
Parts - 100 percent - Labor

How can the RVRN offer such a great warranty?

The procedures followed by RVRN refrigerator specialists have been perfected over a twenty-five-year period. Plus, we do not recondition on an assembly line. Every unit is repaired individually by hand and tested at every stage of the repair to make sure it will last longer and cool better. We know our units are the best, and we back them with the best warranty in the industry.

Compare to others reconditioned cooling units, new cooling units, and new complete refrigerators. Be sure to read their fine print.

www.rvrefrigeration.com

FORD RV REFRIGERATION / RVRN RECONDITIONING PROCEDURES

When troubleshooting has determined that the cooling unit needs to be reconditioned,

it is removed from the cabinet and prepped for repair.

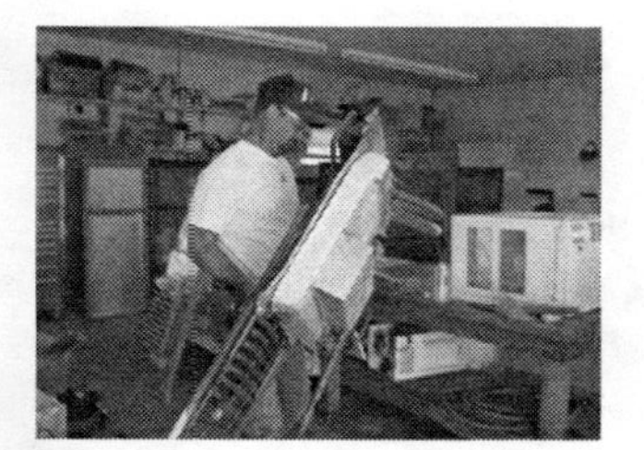

Each cooling unit is repaired individually by a certified technician;

then all repairs are pressure-tested to ensure that not even the smallest leak has been missed.

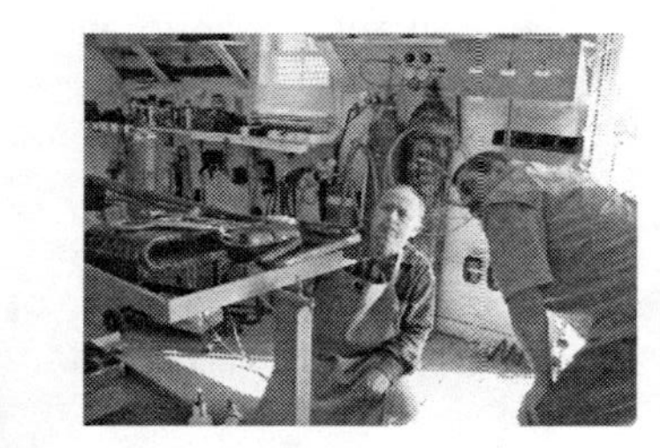

Once the cooling unit is fully repaired, it is charged and fully tested again.

When we are sure the cooling unit is 100 percent, it is reinstalled to the cabinet and retested again.

Step 1
Obtain the Ford procedures manual and familiarize yourself with its contents.

Why is the reconditioning industry booming?

1. Consumers are becoming aware of their options.
2. The RVRN saves consumers money.
3. The RVRN warranty is the best in the industry.

Is there a need for an RVRN refrigerator specialist in your area?

If there is an RV dealer,
an RV service center,
a campground, a resort, or
an Amish community,
the answer is, "*Definitely!*"

Step 2
Schedule training date

- Training is the second week of each month.
- Training is Monday to Friday, 8:00 a.m. to 5:00 p.m. with one hour for lunch.
- A deposit is required to secure the date you selected.
- We understand your situation may change, so once a deposit is made, you have up to one year to obtain the training.

Step 3
Join the RVRN Association
MEMBER BENEFITS

- Best warranty in the industry
- Promotions and referrals
- Newsletter
- Members' certificate
- Catalog discounts
- Networking with others
- Free advertising
- And more

Step 5
Come for training
Training is a hands-on practical and covers the following:

• Theory • Safety • Disassembly • Reconditioning • Recharging • Reassembly • Controls • Marketing • Industry links • Warranties • RVRN • Phone support • And more

Step 4
Secure your accommodations

- Ask about discounted motel rates.
- RV campgrounds are nearby.
- You may park your RV at our facility.
- Restaurants are near the motel and approximately ten minutes from our facility.

Step 6
Purchase customized tool package

QUOTE

"Great people to work with, learned a lot, easy to understand."

—David, service tech in Idaho

THE CUSTOMIZED TOOL PACKAGE

Consists of items from our catalog that are needed to begin reconditioning and recharging RV refrigerator cooling units.

CALL FOR PRICING

(If you qualify for workforce funding, we have a government number.)

Training is at our facility the second week of each month.

Upon completion of training, written and practical exams are given.

CERTIFICATION

Certification will be awarded once you pass both the written and practical exams.

QUOTE
About Training

"I had a very good time here and enjoyed every minute of it. Everyone is friendly and willing to help with all questions. I feel like part of the family."

—Matthew, RV service tech from Utah

GO GREEN-GO RNRN

We service all types of RV's
authorized dealer and service center for the
following:

- A&E
- AAP
- ADCO
- Adler Barbour
- Aqua-Hot
- Aircon Lighting
- Atwood
- Avanti
- Bargman
- Battery Brain
- Blue Sea
- Blue Ox
- Carefree
- Carrier
- Coast
- Coleman
- Deka Batteries
- Demco
- Dimensions
- Dometic
- Duo Therm
- KVH
- King Controls
- Kwikee
- Magic Chef
- Magnum
- Marshall Brass
- MaxxAir
- Norcold
- Outlaw Conversion
- Parallax
- RVW
- Reese
- ROADMASTER
- Suburban
- Tekonsha
- Thetford
- Tripp Lite
- Tundra
- Valterra
- Winegard

RV services include
everything on the RV except the chassis and engine.
Service calls available, or you can bring
your RV to our service center.

1. Used refrigerators
2. Used refrigerator parts
3. Recharging parts catalog

Check our Web site
for more information:
www.rvrefrigeration.com.

NORCOLD TROUBLESHOOTING

SYMPTOMS	POSSIBLE PROBLEM AREAS	CAUSES/REMEDIES **Only for model nos. N51X, N51X.3, N61X, N81X, N62X, N82X, N64X, N84X, N64X.3, N84X.3, N109X, 120XXX**
Nothing is displayed on upper board.	The refrigerator is off. The refrigerator is not connected to the DC volt.	Turn refrigerator on. The vehicle battery is not functioning properly. The AC/DC converter may not be working (if one is installed). **
The letter *a* or no. AC. + alarm	The refrigerator is not receiving voltage from AC.	The refrigerator is unplugged. The fuse or circuit breaker is blown. The generator (if available) is not functioning properly. **

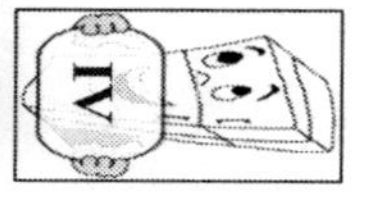
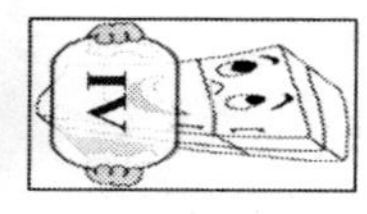

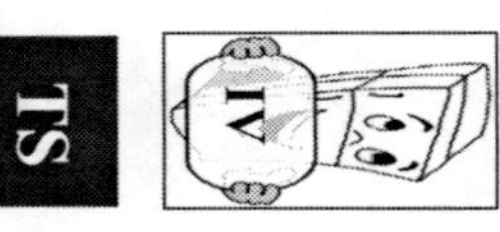

NORCOLD TROUBLESHOOTING

SYMPTOMS	POSSIBLE PROBLEM AREAS	CAUSE / REMEDIES **Only for model #s N51X, N51X.3, N61X, N81X, N62X, N82X, N64X, N84X, N64X.3, N84X.3, N109X, 120XXX**
The letter *C* or DC. LO. + alarm	Insufficient DC voltage.	The battery is not functioning properly. The AC/DC converter may not be working (if one is installed). **
The letter *d* or dr. + alarm	Means the door was left open for over a 2-minute period.	Close the door, and this should cause the electronic board to reset itself.
The letter *E* or dc. HI	The voltage on the DC is too high.	**

NORCOLD TROUBLESHOOTING

SYMPTOMS	POSSIBLE PROBLEM AREAS	CAUSES/REMEDIES (Same model #s as listed on page 71)
The letter *F* or no. FL.	• The burner did not fire up.	• The LP tank is not turned on. • The burner needs to be cleaned. • The LP has the incorrect pressure. • The orifice needs to be cleaned. • The manual shutoff valve on the refrigerator is in the closed position. • There is air in the LP source line (see p. 104, removing air from LP). **
The letter *H*	• Problem with the controls.	**

NORCOLD TROUBLESHOOTING

SYMPTOMS	POSSIBLE PROBLEM AREAS	CAUSES/REMEDIES (Same model #s as listed on page 71)
The letter *n* or no.co + alarm	• Problem with cooling unit function.	• Refrigerator is unlevel (place level in freezer). • Door gasket needs to be replaced. • Poor ventilation from floor to roof on backside of refrigerator. • Heat element is bad or incorrect wattage and needs to be replaced. • Open condensation drain. • The following require a qualified technician: ○ Insufficient insulation between the boiler and cabinet or the flue and the steam line ○ The cooling unit needs to be reconditioned. **

NORCOLD TROUBLESHOOTING

SYMPTOMS	POSSIBLE PROBLEM AREAS	CAUSES/REMEDIES (Same model #s as listed on page 71)
The letter *n* or no.co. + alarm	• Problem with cooling unit sensing	• Heat source to cooling unit have shut down and are locked out by the controls. Turn off the refrigerator by pushing the On/ Off button for two seconds then release. Repeat to turn the unit back on. If the *n* code appears again, the controls will lock and it will require a technician to correct the problem.
AC.HE + alarm	• Problem with heat element	• Heat element is bad or wrong wattage and needs to be replaced.
The letter *r* or AC.rE. + alarm	• Problem with the controls	**

NORCOLD TROUBLESHOOTING

SYMPTOMS	POSSIBLE PROBLEM AREAS	CAUSES/REMEDIES (Same model #s as listed on page 71)
The letter *S* or Sr. + alarm	• Problem with the controls	**
Temp flashes when Temp button pushed	• Backup operating system has kicked in.	**
DC.rE + alarm	• Problem with the controls	**
dc.HE. + alarm	• Problem with the controls	**
AC.LO.	• Problem with the controls	**
AC.HI.	• Problem with the controls	**

NORCOLD TROUBLESHOOTING

SYMPTOMS	POSSIBLE PROBLEM AREAS	CAUSES/REMEDIES (Same model #s as listed on page 71)
I. Will not cool on electric.	1. Incorrect power source	1. Check voltage on refrigerator. a. AC—108-132 volts b. DC—10.5 V minimum for 2-way models, 12 V minimum for 3-way models, 15.4 V maximum
	2. Blown fuse	2. Check fuses with volt-ohm meter. a. Replace if blown with correct size
	3. Thermostat turned to low	3. Set to a higher temperature setting.
	4. Electric—off	4. Set control to electric.
	5. AC/DC switch not set to correct position (3-way models only)	5. Set to correct position.
	6. No voltage at the control board	6. Replace AC/DC switch (3-way models only).

NORCOLD TROUBLESHOOTING

SYMPTOMS	POSSIBLE PROBLEM AREAS	CAUSES/REMEDIES (Same model #s as listed on page 71)
I. Will not cool on electric (continued).	7. Incorrect voltage at the heat element 8. Defective heat element	7. Check for loose wire or replace control. 8. Check for continuity across leads. Check for a short between heat element leads and ground. Replace if open or shorted.
II. Will not cool on gas.	1. LP tanks are empty or have the incorrect pressure. 2. Orifice is dirty.	1. a. Gas pressure should be 11" water column. b. The filter is dirty or clogged. Clean or replace filter. 2. The orifice needs to be cleaned or replaced.

NORCOLD TROUBLESHOOTING

SYMPTOMS	POSSIBLE PROBLEM AREAS	CAUSES/REMEDIES (Same model #s as listed on page 71)
III. Burner flame is yellow or too soft.	1. LP tanks are empty or have the incorrect pressure.	1. a. Gas pressure should be 11" water column. b. The filter is dirty or clogged. Clean or replace filter. 2. The orifice needs to be cleaned or replaced.
IV. Flame either continually goes out or won't light at all.	1. Filter is dirty or clogged. 2. The thermocouple safety valve is out of position or broken. 3. There is improper gas pressure. 4. Orifice is dirty or clogged.	1. Replace control filter. 2. Thermocouple needs to be repositioned. a. Replace thermocouple. 3. Check main line gas pressure. 4. Clean or replace.

NORCOLD TROUBLESHOOTING

SYMPTOMS	POSSIBLE PROBLEM AREAS	CAUSES/REMEDIES (Same model #s as listed on page 71)
V. Flame on burner sounds hard, noisy, or lifting.	1. Gas pressure is too high. 2. Orifice fitting has a gas leak. 3. Orifice is defective or improper. 4. Baffle is missing in flue.	1. Check and reset gas pressure at main line. 2. Leak test the fitting. Using a small amount of dish soap will produce bubbles if there is a leak. Tighten if necessary. 3. Replace orifice. 4. Install baffle (see chart, p. 22 for correct distance).
VI. Burner is hard to light.	1. Gas pressure is low. 2. Electrode igniter is bent.	1. See II on page 77. 2. Adjust electrode igniter (see figure 19, p. 59).

TROUBLESHOOTING THE ELECTRONIC IGNITION

PROBLEM	CAUSE	REMEDY
I. Refrigerator will not light on gas electrically.	1. No 12-V DC 2. Fuse is blown. 3. Ignition switch is off. 4. No 12V to relighter 5. 12-V DC to ignition but no flame	1. Check terminals in rear of refrigerator. 2. Replace fuse. 3. Turn to On position. 4. Replace switch. 5. Check for a. Loose connection b. Broken ceramic c. Electrode in correct position d. Needs to replace relighter

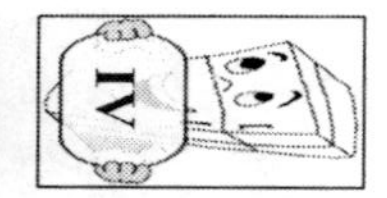

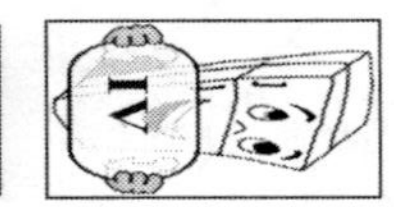

TROUBLESHOOTING THE ELECTRONIC IGNITION

PROBLEM	CAUSE	REMEDY
I. Refrigerator will not light on gas electrically (continued).	6. Flame lights but then goes out.	6. Check electrodes: a. Gas pressure. b. Loosen thermocouple connection at combination control. c. Hold safety valve in all the way. d. Replace thermocouple. e. Replace control.
II. Refrigerator is too cold.	1. Temperature set too high. 2. Ambient temperature extremely cold. 3. Improper installation of capillary tube.	1. Turn down the thermostat. 2. Set thermostat to a warmer setting. 3. Make sure capillary tube is not touching cooling unit and is attached correctly to the fins.

TROUBLESHOOTING THE ELECTRONIC IGNITION

PROBLEM	CAUSE	REMEDY
III. Cooling is not sufficient on gas or electric.	1. Power source is incorrect. 2. Not enough ventilation or incorrect installation. 3. Refrigerator is unlevel. 4. Ammonia odor or yellow coloring on the cooling unit. 5. Air leaking around the doors.	1. Check AC, DC, and gas pressure. 2. Make sure the ventilation is not blocked between the back side of the refrigerator and the outer wall. 3. Level refrigerator by placing a level in the center of the freezer compartment. Check vertical and horizontal. 4. Cooling unit has a leak and needs to be reconditioned. ** 5. Door gasket/s need to be replaced.

TROUBLESHOOTING THE ELECTRONIC IGNITION

PROBLEM	CAUSE	REMEDY
IV. Flame goes out when moving	1. Malfunction with thermocouple. 2. Improper ventilation. 3. Downdrafts.	1. Refer to section IV, page 25. 2. Refer to section III above. 3. Add baffles to roof or contact service tech.

YOUR NOTES

QUOTE

"If nobody builds another new RV again starting today, there are still about 10 million of them on the road, and those units need to be serviced."

—*RV Pro* magazine, February 2009, 4

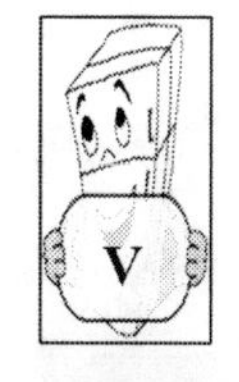

SECTION V

REMOVING THE REFRIGERATOR FROM THE RV

NOTE: Always measure your refrigerator and the exit door of your RV. Sometimes your refrigerator may be too large to fit through the doorway. Your options are (1) to remove the door frame on the RV or (2) to remove a window and its frame. This will have to be done to remove and reinstall the refrigerator.

1. Shut off propane at main tank valve.
2. Go to the refrigerator access door on the outside of the RV. Turn the two locks and lift the access door to see the back side of the refrigerator. (See figure 5-1.)

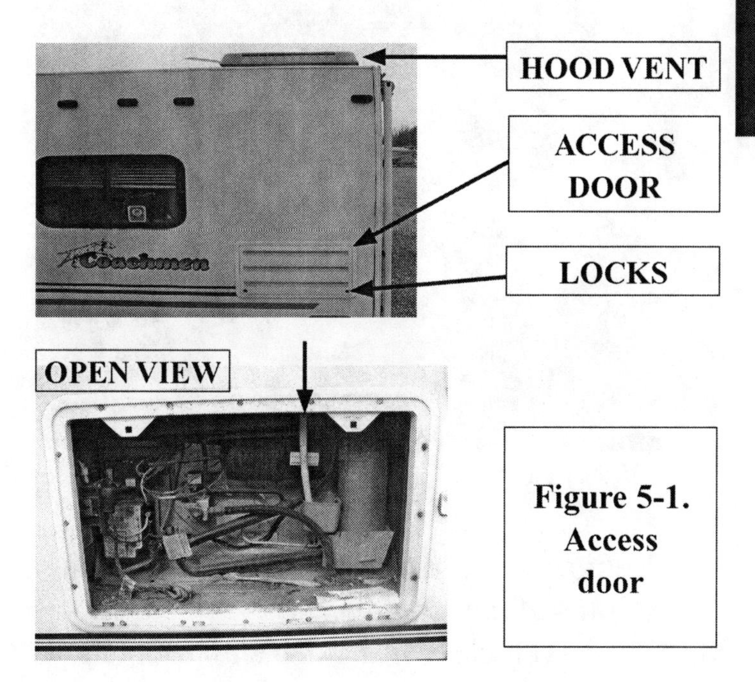

Figure 5-1. Access door

3. Mark positions of 12-volt DC wires (see figure 6-1, page 95).
4. Disconnect 12-volt DC wires and tape positive wire.
5. Disconnect 110-volt AC refrigerator plug (see figure 4-16, page 56).
6. Using two wrenches, one on refrigerator connection and other on brass flare nut, disconnect LP gas line (see figure 5-2). Do not loosen this nut without a backup wrench as indicated. Plug the gas line on the camper in case you use other LP appliances. Make sure it does not leak.

Figure 5-2. Disconnecting LP gas line

7. Remove screws or bolts in lower rear section of unit (if any are present).
8. Inside vehicle, cover floor with drop cloth.

9. Remove refrigerator doors. There are pins or screws that hold the doors on. Just remove the pin or screw out of the hole (see figure 5-3).

Figure 5-3. Door pins

10. Remove screws around frame, which holds the refrigerator to the wall (located at the top, bottom, or sides of refrigerator).
11. Slide the refrigerator out of opening, remove through RV door or window, and take to a service area (this will require assistance from another person). If unit has been in operational mode, be careful of the hot areas, which are the steam line and the boiler.

HOW TO PROPERLY TRANSPORT THE RV REFRIGERATOR

If you plan to take your refrigerator to a technician, use the following procedures to transport the refrigerator.

1. If you and another are carrying the refrigerator, or if you are using a two-wheeler, make sure the hinges of the refrigerator doors are on top so that the doors cannot fall open.

2. Lay the refrigerator down so that the hinges to the doors are on top. Otherwise, the doors can fall open, causing unwanted scratches or other damage. Secure with bungee cords.

QUOTES

"Being in this businesses has been nothing but rewarding. From the feeling you get when you fix your first frig, to the people you meet. I have no regrets about getting started in this hands on business."

—Chris Brain, RVRN member from Iowa

"The manual and training personnel are very professional and prepared. This is the best training I have ever received."

—JR, master certified in HVAC from Florida

SECTION VI
COOLING UNIT DISASSEMBLY

This procedure should only be done if you are replacing the cooling unit. However, remember, the old one can most likely be reconditioned, which will save you money and give you the industry's best warranty if you have an RVRN refrigerator specialist do the reconditioning. In which case, you would not remove the cooling unit; you would take the entire refrigerator or the entire RV to the RVRN refrigerator specialist.

To ensure accurate reassembly, mark positions of any wiring and controls before beginning disassembly (see figure 6-1).

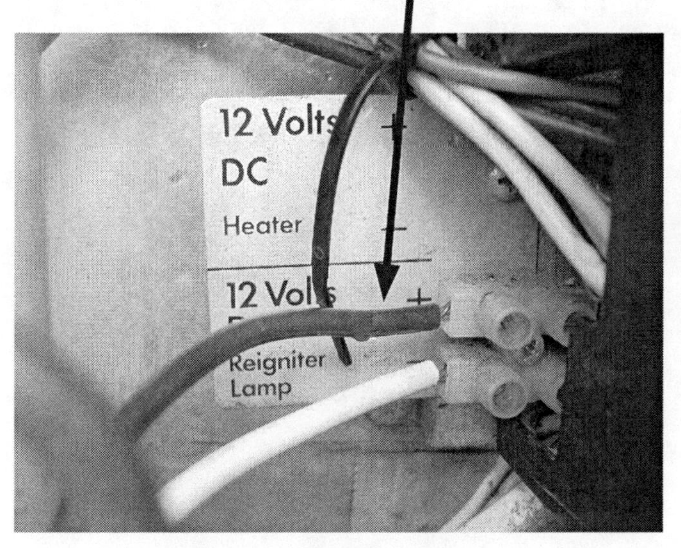

Figure 6-1. Marking wires

1. Remove the screws at the bottom of the cooling unit to make the next procedure easier.
2. Remove the boiler pack and electric heating element(s) (see figure 6-2). (Cooling unit may first need to be removed before this step can be accomplished.)

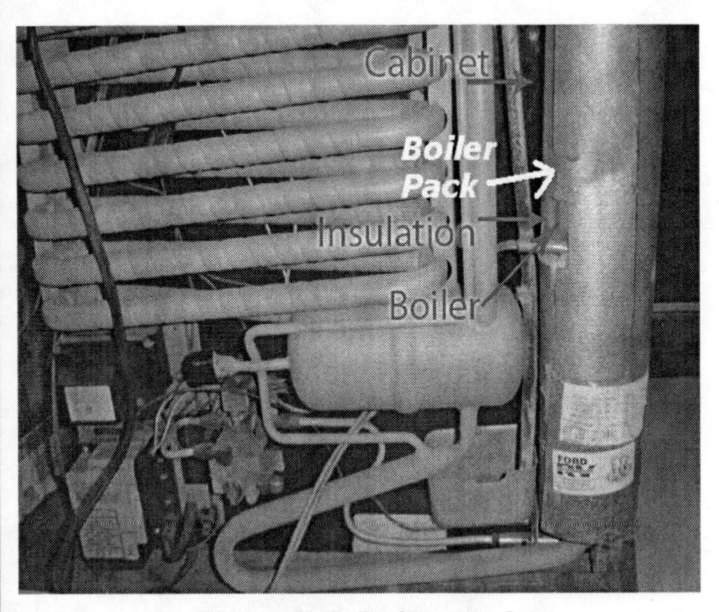

Figure 6-2. Boiler pack

3. Remove the sheet metal cover around burner assembly (see figure 6-3-A and B).

Figure 6-3-A. Cover around burner

Figure 6-3-B. Burner assembly

CAUTION

When accomplishing the following steps, take precautions against crimping the gas line, thermocouple, or thermostat capillary tubes.

4. Remove screws holding burner assembly in place and carefully set burner assembly aside.
5. Remove screws holding cooling unit to box.
6. Extract thermostat capillary tube from inside box if necessary (see figure 6-4).

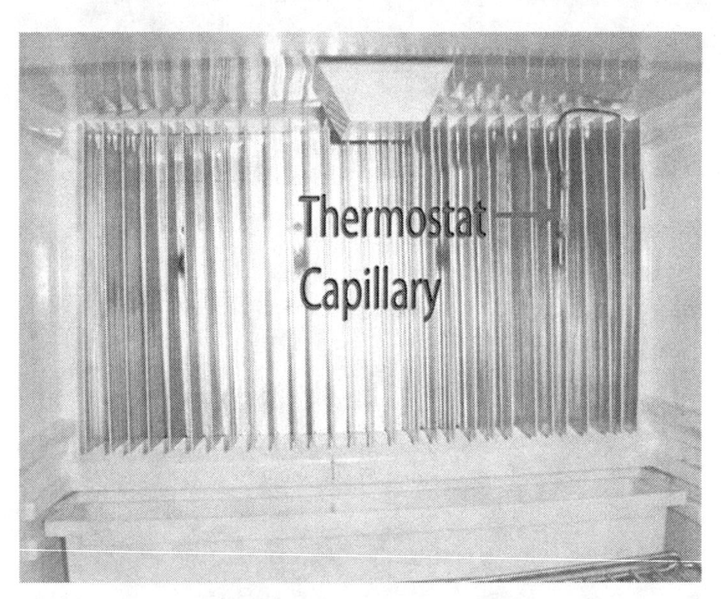

Figure 6-4. Thermostat capillary tube

NOTE

Prior to the removal of the cooling unit, some units require the removal of the cooling fins inside the refrigerator compartment (see figure 6-5) and also the removal of the screws in the freezer (see figure 6-6).

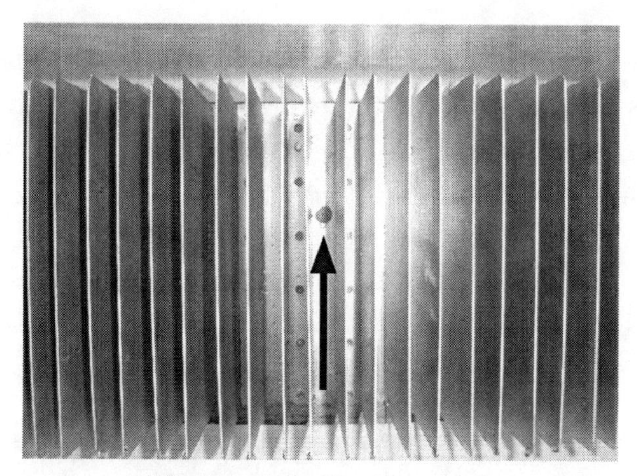

Figure 6-5. Cooling fins

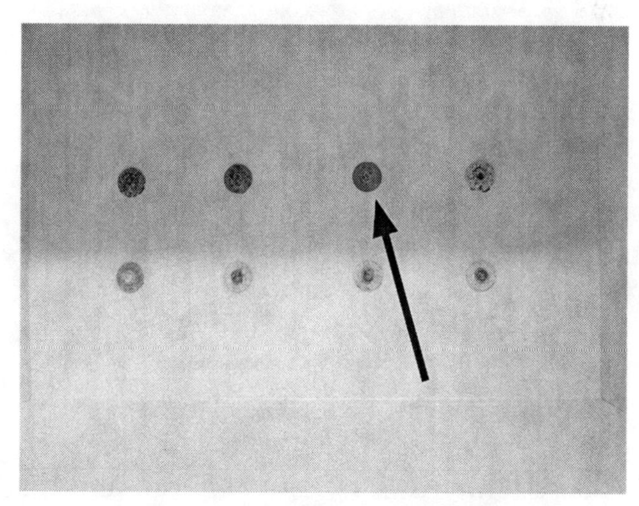

Figure 6-6. Freezer screws

WARNING

The following procedure could be hazardous to eyes. Safety glasses shall be worn at all times. Have eyewash nearby.

CAUTION
BEFORE DOING THIS STEP!

Make sure you have all the screws removed, inside and out. Lift large freezer tube exiting high-temp evaporator. Such force exerted elsewhere could damage unit.

7. Place downward pressure on refrigerator with feet while lifting upward on cooling unit tubing that passes through urethane to and from high-temp evaporator. If a pry bar is necessary, use a scrap piece of wood to protect the box (see figure 6-7).

Figure 6-7. Pry bar and wood

8. The cooling unit is now exposed and ready for replacement.

SECTION VII
COOLING UNIT REASSEMBLY

Boiler Pack

CAUTION

It is critical that you reinstall insulation between the flue tube and the steam line so that they do not touch near the top of the flue tube (see figure 7-1). Be sure to have adequate insulation inside the boiler pack in the area that is between the cabinet and the boiler (see figure 7-2-A and B).

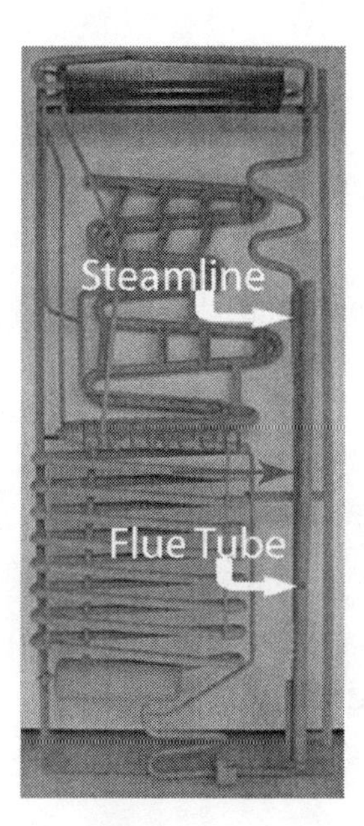

Figure 7-1.
Between steam line and flue tube

Figure 7-2-A and B.
Insulation inside boiler pack

1. Reinstall the heat element and boiler pack around the boiler.

Cooling Unit

2. Accurately reassemble the cooling unit, burner assembly, controls, and wires.
3. When all the gas connections have been made, double-check the connections to make sure none of them are loose. Turn on the gas, and put a mixture of soap and water at the connection. If the soap bubbles, turn off LP and readjust the connection. Operate on gas and leak check again and again until there are no bubbles.

SECTION VIII

FINAL TEST

4. Bypass the controls as shown in procedure 4.2, page 33.
5. Place the thermometer in refrigerator compartment and allow to run overnight (see figure 4-7, page 40). Check the refrigerator temperature the next day. Regardless of ambient temperature, thermometer in refrigerator should read below 32°F.
6. If temperature is satisfactory, reconnect the controls.

7. Allow refrigerator to run 24 hours on electric and 24 hours on gas to ensure that everything is working properly.

YOUR NOTES

QUOTE

"Go Green! Become part of the RV Refrigeration Industry's newest trend by reconditioning RV refrigerator cooling units."

—Roger D. Ford

SECTION IX
HELPFUL HINTS

1. When checking the voltage.
   AC—For proper operation, it should read 101-129 volts. Ideal voltage is 110-120 volts.
   DC—For proper operation, it should read 10.5-15 volts. Ideal voltage is 11-13 volts.
   If these voltages are not present, turn off the refrigerator and contact a certified technician.
2. Removing air from LP line.
   If you have recently refilled your LP tanks, you may have to remove the air trapped in the gas line to your refrigerator.
   a. Switch refrigerator to LP mode.
   b. If it fails to light or if the check light on your refrigerator upper control panel lights up, turn the refrigerator off for approximately 15 seconds and switch refrigerator to LP mode again.
   d. Repeat if necessary.
   e. If unsuccessful after five tries, take to a certified technician.

QUOTE

"Small opportunities are often
the beginning of great successes."

—*Pure Living*, December 2008

SECTION X

REFRIGERATOR MODELS AND DIMENSIONS

DOMETIC REFRIGERATORS

DOMETIC MODEL NUMBER	REFRIGERATOR DIMENSION H × W × D
NDA1402	63 3/16 × 32 3/4 × 26 1/16
NDR1026	59 15/16 × 23 11/16 × 24
NDR1062	59 15/16 × 23 11/16 × 24
NDR1272	59 1/16 × 36 3/4 × 24
NDR1282	59 1/16 × 36 3/4 × 24
NDR1292	59 15/16 × 23 11/16 × 24
NDR1492	59 1/16 × 36 1/2 × 24
RC4000	16 1/4 × 23 5/8 × 21
RM100	51 1/2 × 21 7/8 × 23 3/4
RM1272	59 1/16 × 36 3/4 × 24
RM1282	59 1/16 × 36 3/4 × 24
RM1300-01-03	57 1/2 × 23 3/4 × 23 3/4
RM2111	23 3/4 × 19 3/16 × 18 3/16
RM2150	17 1/4 × 17 3/4 × 20 3/4
RM2190	20 5/8 × 17 1/2 × 21 1/2

DOMETIC MODEL NUMBER	REFRIGERATOR DIMENSION H×W×D
RM2191	21×17 3/4×21 1/2
RM2193	21×17 3/4×21 1/2
RM2202	22 1/4×19 1/4×18 1/4
RM2300-01	29 3/4×20 1/2×21 1/4
RM2310.2-3	29 3/4×20 1/2×21 3/8
RM2332-33	29 3/4×20 1/2×21 3/8
RM2351	29 3/4×20 1/2×21 3/8
RM2354	29 3/4×20 1/2×21 3/8
RM24	22 3/8×19 3/8×19 7/8
RM2400-01	32 3/8×21 7/8×23 3/4
RM2410	32 7/8×22 3/16×24
RM2410.2	32 7/16×22 3/16×24
RM2451	39 9/16×23 11/16×24
RM2452-53	39 9/16×23 11/16×24
RM2454	39 9/16×23 11/16×24
RM2460-61	32 3/8×21 7/8×23 3/4
RM2500-01	40 1/4×21 7/8×23 3/4
RM2510	40 5/16×21 13/16×24
RM2510.2	40 5/16×21 13/16×24
RM2551	42 5/8×23 11/16×24

DOMETIC MODEL NUMBER	REFRIGERATOR DIMENSION H×W×D
RM2552-53	42 5/8×23 11/16×24
RM7732	58 15/16×32 11/16×24
RM77321M	58 15/16×32 11/16×24
RM7832W	58 15/16×32 11/16×24
S620-630	52 5/8×23 1/2×24
S820-30	59 5/8 × 23 13/16 × 24 1/16
S1521-31	42 3/4×23 1/4×24
S1621-31	52 11/16×23 1/4×24
S1821-31	59 1/2×23 1/4×24
T22SRV	21 1/16×18 5/16×221
T80ACRV	53×23 9/16×23 13/16
TF4SRV	34 1/4×22 3/4×22 3/4
RM2554	42 5/8×23 11/16×24
RM2600-01	51 1/2×21 7/8×23 3/4
RM2602-03	46 1/2×21 7/8×23 3/4
RM2604	49 1/2×21 3/4×24
RM2607	49 1/2×21 3/4×24
RM2610	51 1/2×21 3/4×24
RM2611	49 1/2×21 3/4×24
RM2612	49 1/2×21 3/4×24

DOMETIC MODEL NUMBER	REFRIGERATOR DIMENSION H × W × D
RM2620	49 1/2 × 21 3/4 × 24
RM2652	53 3/4 × 23 11/16 × 24
RM2662	53 3/4 × 23 11/16 × 24
RM2800-01	57 1/2 × 23 3/4 × 23 3/4
RM2802-03	64 1/2 × 21 7/8 × 23 3/4
RM2804	53 3/4 × 23 11/16 × 24
RM2810	57 5/16 × 23 3/4 × 24
RM2807 and 2811	55 3/8 × 23 3/4 × 24
RM2812	55 3/8 × 23 5/8 × 24
RM2852	59 15/16 × 23 11/16 × 24
RM2862	59 15/16 × 23 11/16 × 24
RM2880	55 3/8 × 23 5/8 × 24
RM3500-01	38 1/4 × 21 7/8 × 23 3/4
RM36	28 5/8 × 20 1/2 × 20 5/8
RM360-61	29 3/4 × 20 1/2 × 21 1/4
RM3600-01	49 1/2 × 21 7/8 × 23 3/4
RM3607	49 1/2 × 21 3/4 × 24
RM3662-63	53 3/4 × 23 11/16 × 24
RM3800-01	55 1/8 × 23 1/2 × 24
RM3862-63	59 15/16 × 23 11/16 × 24

DOMETIC MODEL NUMBER	REFRIGERATOR DIMENSION H × W × D
RM4223	21 × 19 1/4 × 24
RM45-47	31 1/4 × 21 7/8 × 23 5/8
RM4804	63 × 23 13/16 × 24 1/16
RM4872-73	58 15/16 × 23 11/16 × 24
RM60-66-67	39 1/8 × 21 7/8 × 23 5/8
RM660-63	40 1/2 × 21/78 × 23 3/4
RM75-77	49 1/2 × 21 7/8 × 23 3/4
RM760-63	51 1/2 × 21 7/8 × 23 3/4
RM7732	58 15/16 × 32 11/16 × 24
RM77321M	58 15/16 × 32 11/16 × 24
RM7832W	58 15/16 × 32 11/16 × 24
S620-630	52 5/8 × 23 1/2 × 24
S820-30	59 5/8 × 23 13/16 × 24 1/16
S1521-31	42 3/4 × 23 1/4 × 24
S1621-31	52 11/16 × 23 1/4 × 24
S1821-31	59 1/2 × 23 1/4 × 24
T22SRV	21 1/16 × 18 5/16 × 21
T80ACRV	53 × 23 9/16 × 23 13/16
TF4SRV	34 1/4 × 22 3/4 × 22 3/4

NORCOLD MODEL NUMBER	REFRIGERATOR DIMENSION H ×W ×D
1082	63 1/4 ×32 11/16 ×24
1200LR	63 1/4 ×32 11/16 ×24
322 BG/JB/T	20 5/8 ×17 1/2 ×21 1/4
323TN	30 7/8 ×20 1/2 ×20 1/8
323 BG/JB/T	20 5/8 ×17 1/2 ×21 1/4
332	20 5/8 ×17 1/2 ×21 1/4
3163 t/g/WVG/ HBV	24 1/8 ×25 5/8 ×16 7/8
442-443	30 7/8 ×23 1/2 24
452-453	43 1/4 ×23 1/2 ×24
462-463	52 7/8 ×23 1/2 ×24
482-483	59 7/8 ×23 1/2×24
652-653	43 1/4 ×23 1/2 ×24
662-663	52 7/8 ×23 1/2×24
682-683	59 7/8 ×23 1/2 ×24
6052-6053	43 1/4 ×23 1/2 ×24
6062-6063	53 3/4 ×23 11/16 ×24
6082-6083	59 15/16 ×23 11/16 ×24
6162-6163	52 7/8 ×23 1/2 ×24

NORCOLD MODEL NUMBER	REFRIGERATOR DIMENSION H × W × D
6182-6183	59 7/8 × 23 1/2 × 24
703-704, 707	59 7/8 × 23 1/2 × 24
723, 725-727, 743, 747	59 7/8 × 23 1/2 × 24
774	33 1/8 × 22 × 21 1/2
776	40 1/8 × 22 × 21 1/2
778	52 1/8 × 23 7/8 × 24
838	52 5/8 × 22 × 21 3/4
874	33 × 22 × 21 3/4
874EG2-874EG3	30 7/8 × 23 1/2 × 24
875	40 × 22 × 21 3/4
875EG2-875EG3	39 7/8 × 22 × 21 11/16
876	52 1/8 × 24 × 24
876EG2	52 7/8 × 23 15/16 × 24
878	59 × 23 15/16 × 24
878EG2-878EG3	59 7/8 × 23 15/16 × 24
8010	59 × 23 7/8 × 24
8010EG2-8010EG3	59 7/8 × 23 1/2 × 24
8310	58 7/8 × 23 15/16 × 23 1/2

NORCOLD MODEL NUMBER	REFRIGERATOR DIMENSION H × W × D
8310EG2-8310EG3	59 7/8 × 23 1/2 × 24
8652-8653	43 × 23 5/8 × 24
8662-8663	52 7/8 × 23 11/16 × 24
8682-8283	59 5/16 × 23 11/16 × 24
962, 962IM	52 7/8 × 23 1/2 × 24
963, 963IM	52 7/8 × 23 1/2 × 24
982, 982IM	59 7/8 × 23 1/2 × 24
983, 983IM	52 7/8 × 23 1/2 × 24
9162-9163	52 7/8 × 23 1/2 × 24
9182-9183	59 7/8 × 23 1/2 × 24
N260	21 × 19 1/4 × 24
N300 Series	29 3/4 × 20 1/2 × 21 3/8
N510 Series	42 5/8 × 23 11/16 × 24
N641 Series	52 7/8 × 23 1/2 × 24
N821 Series	59 15/16 × 23 11/16 × 24
N841 Series	59 7/8 × 23 15/16 × 24
N1095 Series	59 7/8 × 23 1/2 × 24
1200 LR Series	63 1/4 × 32 11/16 × 24
1200AC Only	63 1/4 × 32 11/16 × 24

INSTAMATIC MODEL NUMBER	REFRIGERATOR DIMENSION H × W × D
RV–2–201	21 3/4 × 19 3/16 × 18 3/16
RV–3–301	28 3/4 × 20 11/16 × 19 7/16
RV–4–401	33 15/16 × 20 11/16 × 19 7/16
RV–6–601	40 3/16 × 22 5/16 × 22 1/8
RV–7–701	53 1/8 × 23 7/8 × 23 1/4
RV 8	59 5/16 × 23 7/8 × 23 3/4
IM 2	22 1/4 × 18 1/8 × 18 5/8
IM 3	28 3/8 × 20 1/4 × 19 5/8
IM 4	31 × 21 5/8 × 22 5/8
IM 6	38 7/8 × 21 5/8 × 22 5/8
IM 7	56 1/2 × 23 1/2 × 22 3/8
IM 10	55 1/8 × 23 1/2 × 225/8
IM 40	31 1/2 × 24 × 23
IM 60	40 1/2 × 24 × 23
M 70	52 × 24 × 23
IM 140	31 1/4 × 22 × 22 3/4
IM 160	40 1/4 × 22 × 22 3/4
M 270	51 3/4 × 23 3/4 × 23 1/4

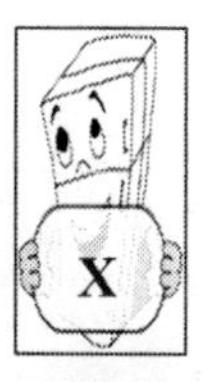

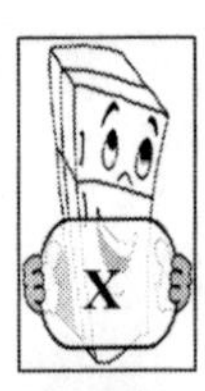

GE REFRIGERATORS

GE MODEL NUMBER	REFRIGERATOR DIMENSION
TG3	28 3/8 × 20 1/4 × 19 5/8
TG4	31 × 21 5/8 × 22 5/8
TG5	38 7/8 × 21 5/8 × 22 5/8

SIBIR REFRIGERATORS

SIBIR MODEL NUMBER	REFRIGERATOR DIMENSION
RV104	30 5/16 × 23 1/2 × 23
RV106	34 3/8 × 23 1/2 × 23
RV108	54 3/8 × 23 1/2 × 23
RV110	64 3/8 × 23 1/2 × 23
RV240	34 1/8 × 23 1/2 × 23
RV270	54 1/16 × 23 1/2 × 23
RV280	59 × 23 7/8 × 24

HADCO REFRIGERATORS

HADCO MODEL NUMBER	REFRIGERATOR DIMENSION
HR2	21 3/4 × 19 3/16 × 18 3/16
HR3	28 3/8 × 20 1/4 × 19 5/8
HR4	31 × 21 5/8 × 22 5/8
HR6	38 7/8 × 21 5/8 × 22 5/8
HR7	56 1/2 × 23 1/2 × 23 3/8

MAGIC CHEF REFRIGERATORS

MAGIC CHEF MODEL NUMBER	REFRIGERATOR DIMENSION
TR–82	56 1/2 × 23 1/2 × 22 3/4
R–87	56 1/2 × 23 1/2 × 23 3/8
R– 88	51 1/4 × 21 5/8 × 22 3/4

YOUR NOTES

QUOTE

"This was the best training I have ever attended. I highly recommend, especially to an HVAC company."

—Chauncey, HVAC master certified and HVAC company owner from Arizona, 2009

SECTION XI
HEAT ELEMENT WATTAGE AND VOLTAGE

DOMETIC REFRIGERATORS

* Indicates that two heat elements are required

To ensure the correct heat element, remove the old heat element and check the information on it.

Remove the heat element leads from the lower circuit board to measure continuity.

DOMETIC MODEL NUMBER	AC HEATING ELEMENT		DC HEATING ELEMENT	
	WATTS +/– 10%	VOLTS	WATTS +/– 10%	VOLTS
FC140	160	120	160	12
M28, C, D	135	110	–	–
M70	225	110	–	–
MA35	135	110	–	–
MA40C	135 225	110 110	– –	– –
MB52C	225	110	–	–
NDR1272	210	120	–	–
NDR1282	210	120	–	–

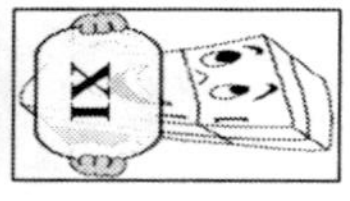

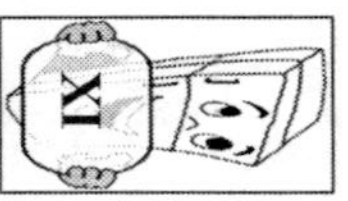

DOMETIC MODEL NUMBER	AC HEATING ELEMENT		DC HEATING ELEMENT	
	WATTS +/– 10%	VOLTS	WATTS +/– 10%	VOLTS
NDR1292	210	120	–	–
NDR1492	210	120	–	–
RA1302	325	120	–	–
RC150	60	120	60	12
RC150TEG	–	–	85	12
RC152	65	110	65	110/12
RC160E	–	–	85	12
RC160EGP	–	–	85	12
RC1600	75	–	85	12

DOMETIC MODEL NUMBER	AC HEATING ELEMENT		DC HEATING ELEMENT	
	WATTS +/– 10%	VOLTS	WATTS +/– 10%	VOLTS
RC1600EGP	–	–	85	12
RC2000	75	–	85	–
RC36D	135	110	135	110/12
RM17A	95	110	–	–
RM18	65	110	65	110/12
RM23	95	110	–	–
RM24	95	110	95	110/12
RM36	135	110	135	110/12
RM36A	135	110	135	110/12

DOMETIC MODEL NUMBER	AC HEATING ELEMENT		DC HEATING ELEMENT	
	WATTS +/– 10%	VOLTS	WATTS +/– 10%	VOLTS
RM36C	135	110	135	110/12
RM36D	135	110	–	–
RM36E	135	110	135	12
RM45	135	110	135	110/12
RM45A	135	110	–	–
RM46	135	110	135	110/12
RM46D	135	110	135	110/12
RM46E	135	110	–	–
RM47	135	110	135	110/12

DOMETIC MODEL NUMBER	AC HEATING ELEMENT		DC HEATING ELEMENT	
	WATTS +/– 10%	VOLTS	WATTS +/– 10%	VOLTS
RM47D	135	110	135	110/12
RM60	150	110	135	110/12
RM60A	150	110	135	110/12
RM66	150	110	135	110/12
RM66D	135	110/12	–	–
RM66E/F	160	120	160	12
RM67	150	110	135	110/12
RM67D	175	110	135	110/12
RM75	225	110	–	–

DOMETIC MODEL NUMBER	AC HEATING ELEMENT		DC HEATING ELEMENT	
	WATTS +/– 10%	VOLTS	WATTS +/– 10%	VOLTS
RM75A	225	110	–	–
RM76	225	110	–	–
RM76D	225	110	–	–
RM77	225	110	–	–
RM77D	225	110	–	–
RM100	275	110/120	275	12
RM182	85	110	85	12
RM182AGA	85	110	85	12
RM182B	85	120	85	12

DOMETIC MODEL NUMBER	AC HEATING ELEMENT		DC HEATING ELEMENT	
	WATTS +/– 10%	VOLTS	WATTS +/– 10%	VOLTS
RM182BEGI	85	120	85	12
RM190	95	110/12	–	–
RM211	95	110/12	–	–
RM211B	95	110/12	–	–
RM215	–	–	95	12
RM235	95	110/12	–	–
RM360	150	120	125	12
RM361	150	120	125	12
RM460	150	120	125	12

DOMETIC MODEL NUMBER	AC HEATING ELEMENT		DC HEATING ELEMENT	
	WATTS +/– 10%	VOLTS	WATTS +/– 10%	VOLTS
RM461	160	120	125	12
RM660	175	115	175	12
RM661	175	115	175	12
RM663	210	120	175	12
RM760	295	120	250	12
RM761	295	120	250	12
RM763	295	120	260	12
RM1272	210	120	275	12
RM1282	210	120	275	12

DOMETIC MODEL NUMBER	AC HEATING ELEMENT		DC HEATING ELEMENT	
	WATTS +/– 10%	VOLTS	WATTS +/– 10%	VOLTS
RM1300-01-03	325	120	275	12
RM2150	–	–	95	12
RM2190	–	–	95	12
RM2191	–	–	115	12
RM2192	95	110	95	12
RM2193	115	110	115	12
RM2200a	95	110/12	–	–
RM2201	95	110	95	12
RM2202	125	–	125	12

DOMETIC MODEL NUMBER	AC HEATING ELEMENT		DC HEATING ELEMENT	
	WATTS +/– 10%	VOLTS	WATTS +/– 10%	VOLTS
RM2300-01	160	120	125	12
RM2310	160	120	125	12
RM2332	175	120	125	12
RM2410	160	–	125	–
RM2452	210	–	–	–
RM2453	210	–	175	–
RM2510	185	–	175	–
RM2552	210	–	–	–
RM2553	210	–	175	–

DOMETIC MODEL NUMBER	AC HEATING ELEMENT		DC HEATING ELEMENT	
	WATTS +/– 10%	VOLTS	WATTS +/– 10%	VOLTS
RM2604	295	120	215	–
RM2607	295	120	215	12
RM2610	295	120	215	–
RM2611	295	120	215	12
RM2612	295	120	–	–
RM2620	325	110	–	–
RM2652	325	110	–	–
RM2800	325	120	275	12
RM2801	325	120	215	12

DOMETIC MODEL NUMBER	AC HEATING ELEMENT		DC HEATING ELEMENT	
	WATTS +/– 10%	VOLTS	WATTS +/– 10%	VOLTS
RM2802-03	325	120	–	–
RM2804	325	120	215	12
RM2807	325	120	215	12
RM2810	325	120	–	–
RM2811	325	120	215	12
RM2812	325	110	–	–
RM2820	325	110	–	–
RM2852	325	110	–	–
RM3500-01	185	120	175	12

DOMETIC MODEL NUMBER	AC HEATING ELEMENT		DC HEATING ELEMENT	
	WATTS +/– 10%	VOLTS	WATTS +/– 10%	VOLTS
RM3600	295	120	250	12
RM3601	295	120	215	12
RM3604	295	120	215	12
RM3607	295	120	215	12
RM3662	325	110	–	–
RM3663	325	120	215	12
RM3800	325	120	275	12
RM3801	325	120	215	12
RM3804	325	120	215	12

DOMETIC MODEL NUMBER	AC HEATING ELEMENT		DC HEATING ELEMENT	
	WATTS +/– 10%	VOLTS	WATTS +/– 10%	VOLTS
RM3807	325	120	215	12
RM3862	325	120	–	–
RM3863	325	120	215	12
RM4223	125	120	–	–
RM4801	325	120	215	12
RM4804	325	120	215	12
RM4804.002	325	120	215	12
RM4804.004	325	120	215	12
RM4804.005	325	120	215	12

DOMETIC MODEL NUMBER	AC HEATING ELEMENT		DC HEATING ELEMENT	
	WATTS +/– 10%	VOLTS	WATTS +/– 10%	VOLTS
RM4804.006	325	120	215	12
RM4872	325	120	–	–
RM4873	325	120	215	12
RM7030	210*	120	–	–
RM7130	210*	120	–	–
RM7732	210*	120	–	–
RM7832W	210*	120	–	–
S1521	195	120	–	–
S1531	185	120	215	12

DOMETIC MODEL NUMBER	AC HEATING ELEMENT		DC HEATING ELEMENT	
	WATTS +/– 10%	VOLTS	WATTS +/– 10%	VOLTS
S1621	325	120	215	12
S1631	325	120	215	12
S1821	325	120	–	–
S1831	325	120	215	–
S520.012	170	–	–	–
S520.013	170	–	–	–
S520.014	185	–	–	–
S520.015	185	–	–	–
S530.012	170	–	150	–

DOMETIC MODEL NUMBER	AC HEATING ELEMENT		DC HEATING ELEMENT	
	WATTS +/– 10%	VOLTS	WATTS +/– 10%	VOLTS
S530.013	170	–	15	–
S619.006	300	–	–	–
S619.007	300	–	–	–
S620.012	300	–	–	–
S620.013	300	–	–	–
S620.014	325	–	–	–
S620.015	325	–	–	–
S630.012	300	–	225	–
S630.013	300	–	225	–

DOMETIC MODEL NUMBER	AC HEATING ELEMENT		DC HEATING ELEMENT	
	WATTS +/– 10%	VOLTS	WATTS +/– 10%	VOLTS
S630.014	325	–	215	–
S630.015	325	–	215	–
S819.006	300	–	–	–
S819.007	300	–	–	–
S820.012	300	–	–	–
S820.013	300	–	–	–
S820.014	325	–	215	–
S820.015	325	–	215	–
S830.012	300	–	215	–

DOMETIC MODEL NUMBER	AC HEATING ELEMENT		DC HEATING ELEMENT	
	WATTS +/– 10%	VOLTS	WATTS +/– 10%	VOLTS
S830.013	300	–	225	–
S830.014	325	–	215	–
S830.015	325	–	215	–

Requirements for DC Voltage

- DC voltage can vary up to 22V. However, if voltage goes over the maximum, the unit will shut down automatically, allowing the DC voltage to reduce down to 18V DC. If the voltage drops below 10.5V, this may interfere with the function of the refrigerator controls.
- Panel lights may remain functional unless voltage drops to 4V DC or less.
- Proper polarity is critical. Never use the chassis or body as a conductor.

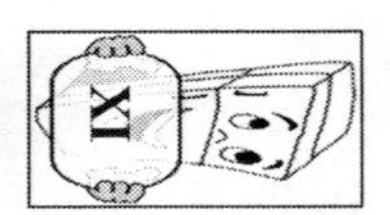

NORCOLD MODEL NUMBER	AC HEATING ELEMENT		DC HEATING ELEMENT	
	WATTS +/– 10%	VOLTS	WATTS +/– 10%	VOLTS
1082	300–350	110	–	–
1200LR	225	110	–	–
322 BG/JB/T	140	110	–	–
323TN	140	110	–	–
323 BG/JB/T	140	110	–	–
332	140	110	–	–
3163 t/g/WVG/HBV	140	110	140	12
442	170	110	–	–
443	170	110	150	12

NORCOLD MODEL NUMBER	AC HEATING ELEMENT		DC HEATING ELEMENT	
	WATTS +/– 10%	VOLTS	WATTS +/– 10%	VOLTS
452-453	170	110	150	12
462	300/350	110	–	–
463	300	110	225	12
482	300/350	110	–	–
483	300	110	225	12
624	250	110	–	–
643	135	110	–	–
644	200	110	–	–
646	250	110	–	–

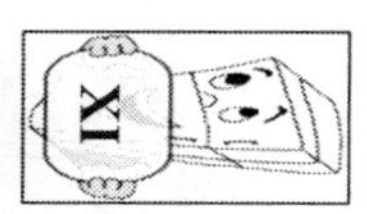

NORCOLD MODEL NUMBER	AC HEATING ELEMENT		DC HEATING ELEMENT	
	WATTS +/– 10%	VOLTS	WATTS +/– 10%	VOLTS
647	250	110	–	–
652-653	170	110	150	12
662-663	300	110	225	12
682/682IM	300	110	–	–
683	135	110	225	12
6052	200	110	–	–
6053	170/200	110	225	12
6062	300	110	–	–
6063	300	110	150	12

NORCOLD MODEL NUMBER	AC HEATING ELEMENT		DC HEATING ELEMENT	
	WATTS +/– 10%	VOLTS	WATTS +/– 10%	VOLTS
6082	300	110	–	–
6083	300	110	225	12
6162	300	110	–	–
6163	300	110	150	12
6182	300	110	–	–
6183	300	110	150	12
703	135	110	–	–
704	200	110	–	–
707	250	110	–	–

NORCOLD MODEL NUMBER	AC HEATING ELEMENT		DC HEATING ELEMENT	
	WATTS +/– 10%	VOLTS	WATTS +/– 10%	VOLTS
723	250	110	–	–
774	200	110	–	–
725	200	110	–	–
727	250	110	–	–
743	250	110	–	–
747	250	110	–	–
774	200	110	–	–
*774	200	110	150	12
776	200	110	–	–

NORCOLD MODEL NUMBER	AC HEATING ELEMENT		DC HEATING ELEMENT	
	WATTS +/– 10%	VOLTS	WATTS +/– 10%	VOLTS
*776	200	110	150	12
778	200	110	–	–
778 EG	350	110	250	12
838	350	110	–	–
*838	350	110	225	12
838EG2	350	110	–	–
838EG3	350	110	225	12
865	170	110	150	12
866	–	–	225	12

NORCOLD MODEL NUMBER	AC HEATING ELEMENT		DC HEATING ELEMENT	
	WATTS +/– 10%	VOLTS	WATTS +/– 10%	VOLTS
868	–	–	225	12
874	–	–	150	12
874EG2	170	110	–	–
874EG3	–	–	150	12
875	–	–	150	12
875EG2	170	110	150	14
875EG3	–	–	150	14
876	–	–	225	12
876EG2	300	110	225	12

NORCOLD MODEL NUMBER	AC HEATING ELEMENT		DC HEATING ELEMENT	
	WATTS +/– 10%	VOLTS	WATTS +/– 10%	VOLTS
878	–	–	225	12
903	–	–	225	12
963	–	–	225	12
983	–	–	225	12
Standard burner	**Standard burner**	**Standard burner**	**Standard burner**	**Standard burner**
876EG2	350/225	120/12	150	12
878EG2	350/300	120/110	–	–
878EG3	350/225	120/12	225	12

NORCOLD MODEL NUMBER	AC HEATING ELEMENT		DC HEATING ELEMENT	
	WATTS +/– 10%	VOLTS	WATTS +/– 10%	VOLTS
8010 EG2	350	120	–	–
8010EG3	350	120	–	–
8310EG2	350	120	–	–
8310EG3	350	110	225	12
8652	170/200	110	–	–
8653	–	–	150	12
962/962IM	300	110	–	–
963/963IM	–	–	225	12
982/982IM	300	110	–	–

NORCOLD MODEL NUMBER	AC HEATING ELEMENT		DC HEATING ELEMENT	
	WATTS +/– 10%	VOLTS	WATTS +/– 10%	VOLTS
983/983IM	–	–	225	12
9162	300	110	–	–
9163	300	110	225	12
9182	300	110	–	–
9183	350	110	225	12
N260	140	110	–	–
N2603	–	–	140	12
N300	180	120	–	–
N3003	180	120	–	–

NORCOLD MODEL NUMBER	AC HEATING ELEMENT		DC HEATING ELEMENT	
	WATTS +/– 10%	VOLTS	WATTS +/– 10%	VOLTS
N641/IM	300	110	–	–
N6413	–	–	225	12
N821	300	110	–	–
N841	300	110	225	12

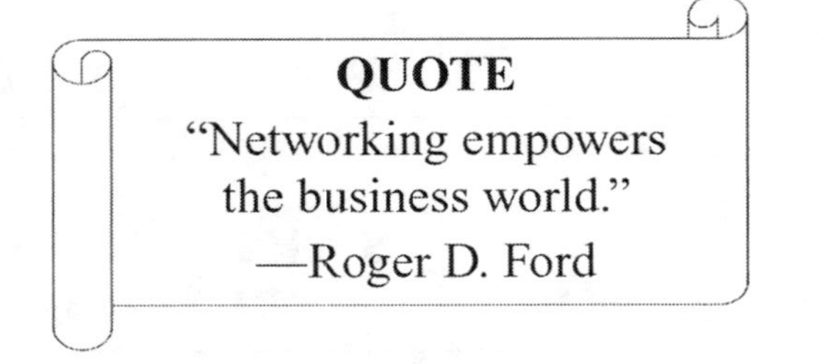

YOUR NOTES

INSTAMATIC MODEL NUMBER	AC HEATING ELEMENT		DC HEATING ELEMENT	
	WATTS +/– 10%	VOLTS	WATTS +/– 10%	VOLTS
RV2	110	110	110	12
RV3	140	110	135	12
RV4	170	110	170	12
RV6	200	110	–	–
RV7	280	110	–	–
RV8	340	110	–	–
RV55 (straight)	250	110	250	12

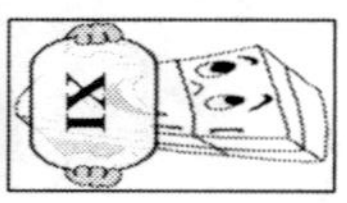

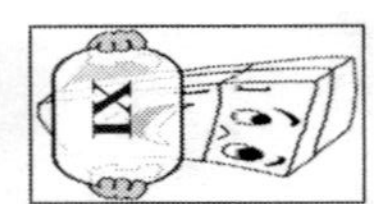

INSTAMATIC MODEL NUMBER	AC HEATING ELEMENT		DC HEATING ELEMENT	
	WATTS +/– 10%	VOLTS	WATTS +/– 10%	VOLTS
IM2	95	110	95	12
IM3	120	110	135	12
IM4	140	110	160	12
IM6	160	110	160	12
IM7	225	110	–	–
IM10 (pencil)	275	110	275	12
IM12	325	110	275	12

INSTAMATIC MODEL NUMBER	AC HEATING ELEMENT		DC HEATING ELEMENT	
	WATTS +/– 10%	VOLTS	WATTS +/– 10%	VOLTS
IM42 (straight)	170	110	–	–
IM42 (pencil 90°)	170	110	–	–
IM43 (straight)	170	110	220	12
IM43 (pencil 90°)	170	110	170	12
IM62 (straight)	250	110	–	–
IM62 (pencil 90°)	220	110	–	–
IM63 (straight)	250	110	250	12

INSTAMATIC MODEL NUMBER	AC HEATING ELEMENT		DC HEATING ELEMENT	
	WATTS +/– 10%	VOLTS	WATTS +/– 10%	VOLTS
IM63 (pencil 90°)	220	110	220	12
IM72 (straight)	250	110	–	–
IM72 (pencil)	260	110	–	–
IM72 (slight bend)	220	110	–	–
IM73 (straight)	250	110	250	12
IM73 (pencil)	260	110	260	12

INSTAMATIC MODEL NUMBER	AC HEATING ELEMENT		DC HEATING ELEMENT	
	WATTS +/– 10%	VOLTS	WATTS +/– 10%	VOLTS
IM73 (slight bend)	220	110	220	12
IM92 (straight)	250	110	–	–
IM92 (pencil 90°)	260	110	–	–
IM92 (slight bend)	220	110	–	–
IM93 (straight)	250	110	–	–
IM93 (pencil 90°)	260	110	260	12

INSTAMATIC MODEL NUMBER	AC HEATING ELEMENT		DC HEATING ELEMENT	
	WATTS +/– 10%	VOLTS	WATTS +/– 10%	VOLTS
IM93 (slight bend)	220	110	220	12
IM142 (straight)	170	110	–	–
IM143 (straight)	170	110	170	12
IM142 (pencil 90°)	170	110	–	–
IM143 (pencil 90°)	170	110	–	–
IM162 (straight)	250	110	–	–

INSTAMATIC MODEL NUMBER	AC HEATING ELEMENT		DC HEATING ELEMENT	
	WATTS +/– 10%	VOLTS	WATTS +/– 10%	VOLTS
IM162 (straight)	250	110	–	–
IM163 (straight)	250	110	–	–
IM163 (straight)	250	110	–	–
IM272 (straight)	250	110	–	–
IM273 (straight)	250	110	–	–
IM292 (straight)	250	110	–	–

INSTAMATIC MODEL NUMBER	AC HEATING ELEMENT		DC HEATING ELEMENT	
	WATTS +/– 10%	VOLTS	WATTS +/– 10%	VOLTS
IM293 (straight)	250	110	–	–

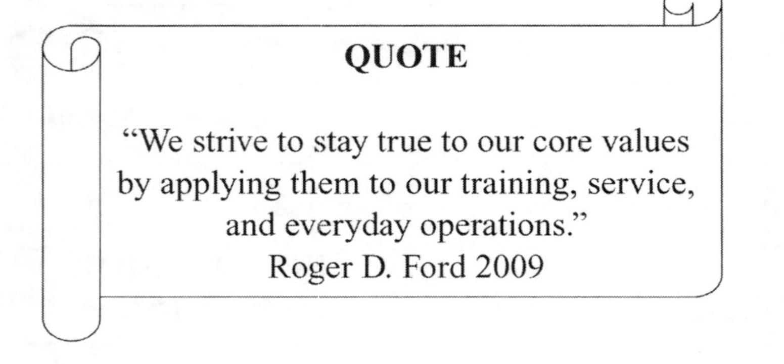

QUOTE

"We strive to stay true to our core values by applying them to our training, service, and everyday operations."
Roger D. Ford 2009

YOUR NOTES

GE MODEL NUMBER	AC HEATING ELEMENT WATTS +/– 10%	AC HEATING ELEMENT VOLTS	DC HEATING ELEMENT WATTS +/– 10%	DC HEATING ELEMENT VOLTS
TG3	135	110	–	–
TG4	150	110	–	–
TG5	150	110	–	–
TG7	225	110	–	–

QUOTE

"Ongoing education keeps your mind stronger and your piggy bank fuller."
—Onna Lee Ford

YOUR NOTES

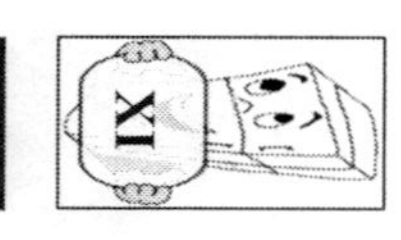

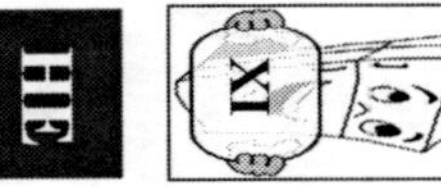

HADCO MODEL NUMBER	AC HEATING ELEMENT		DC HEATING ELEMENT	
	WATTS +/– 10%	VOLTS	WATTS +/– 10%	VOLTS
HR2	95	110	95	12
HR3	135	110	–	–
HR4	150	120	–	–
HR6	175	120	–	–
HR7	225	110	–	–
MKM	135	110	–	–
MKM110	135	110	–	–
300	100	110	–	–
410	125	110	–	–
500	125	110	–	–

HADCO MODEL NUMBER	AC HEATING ELEMENT		DC HEATING ELEMENT	
	WATTS +/– 10%	VOLTS	WATTS +/– 10%	VOLTS
600	125	110	–	–
700	90/60	110	–	–

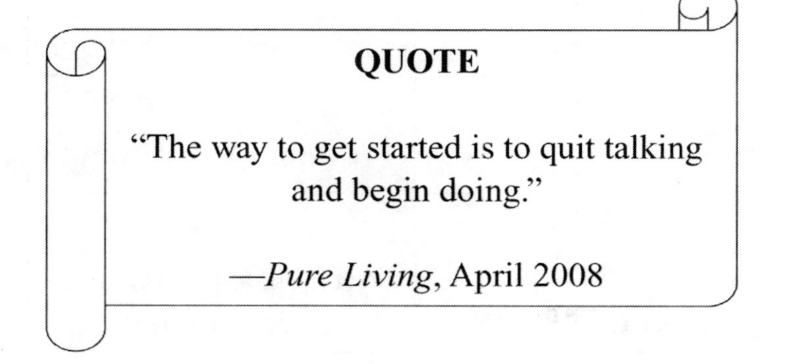

QUOTE

"The way to get started is to quit talking and begin doing."

—*Pure Living*, April 2008

YOUR NOTES

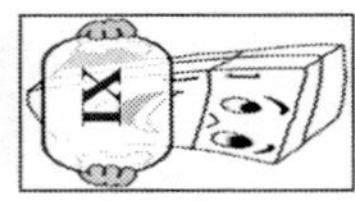

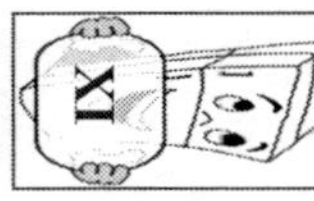

FRIGKING MODEL NUMBER	AC HEATING ELEMENT		DC HEATING ELEMENT	
	WATTS +/– 10%	VOLTS	WATTS +/– 10%	VOLTS
7306DO	115	110	115	12
7403DO	140	110	–	–
7404DO	140	110	–	–
73063	115	110	–	–
74033	140	110	–	–
74043	140	110	–	–
7306DDO	115	110	115	12
7403DDO	140	110	140	12
7404DDO	140	110	140	12

FRIGKING MODEL NUMBER	AC HEATING ELEMENT WATTS +/– 10%	AC HEATING ELEMENT VOLTS	DC HEATING ELEMENT WATTS +/– 10%	DC HEATING ELEMENT VOLTS
Dual Wattage	**Dual Wattage**	**Dual Wattage**	**Dual Wattage**	**Dual Wattage**
7600	60/105	110	60/105	12
700	90/60	110	–	–

YOUR NOTES

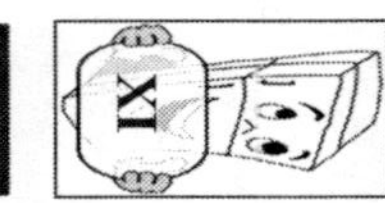
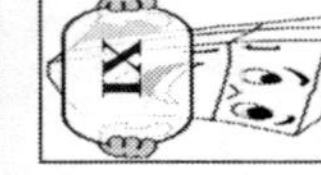

WINNEBAGO MODEL NUMBER	AC HEATING ELEMENT		DC HEATING ELEMENT	
	WATTS +/– 10%	VOLTS	WATTS +/– 10%	VOLTS
W4	170	110	–	–
W6	200	110	–	–
W7	280	110	–	–
W8	340	110	–	–
E1221/2A	160/65	120	–	–
E1221/3A	160/65	120	–	–

MAGIC CHEF MODEL NUMBER	AC HEATING ELEMENT		DC HEATING ELEMENT	
	WATTS +/– 10%	VOLTS	WATTS +/– 10%	VOLTS
MKM	225	110	–	–
M40	135	110	–	–
R30	168	110	–	–
R31	168	110	–	–
R35	168	110	–	–
	170	110/12	–	–
R36	168	110	–	–
	170	110/12	–	–
R37	115	110	–	–

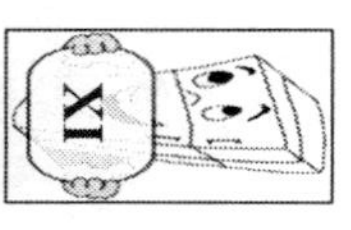

MAGIC CHEF MODEL NUMBER	AC HEATING ELEMENT		DC HEATING ELEMENT	
	WATTS +/– 10%	VOLTS	WATTS +/– 10%	VOLTS
R38	115	110	–	–
R40	50/115	110	–	–
R41	50/115	110	–	–
R44	135	110	–	–
	115/135	110/12	–	–
R45	135	110	–	–
	115/135	110/12	–	–
R46	135	110	–	–
	115/135	110/12	–	–

MAGIC CHEF MODEL NUMBER	AC HEATING ELEMENT		DC HEATING ELEMENT	
	WATTS +/– 10%	VOLTS	WATTS +/– 10%	VOLTS
R47	135	110	–	–
	115/135	110/12	–	–
R48	135	110	–	–
	115/135	110/12	–	–
R50	50/135	110	–	–
R51	50/135	110	–	–
R54	135	110	–	–
R55	135	110	–	–
	115/135	110/12	–	–

MAGIC CHEF MODEL NUMBER	AC HEATING ELEMENT WATTS +/– 10%	AC HEATING ELEMENT VOLTS	DC HEATING ELEMENT WATTS +/– 10%	DC HEATING ELEMENT VOLTS
R56	135	110	—	—
	115/135	110/12	—	—
R57	135	110	—	—
	115/135	110/12	—	—
R58	135	110	—	—
	115/135	110/12	—	—
R65/66/67	135	110	—	—
	115/135	110/12	—	—
R68	135	110	—	—

MAGIC CHEF MODEL NUMBER	AC HEATING ELEMENT		DC HEATING ELEMENT	
	WATTS +/– 10%	VOLTS	WATTS +/– 10%	VOLTS
*R82	225	110	–	–
R82	220	110	–	–
*R83	225	110	–	–
R83	220	110	–	–
*R84	225	110	–	–
R84	220	110	–	–
*R85	225	110	–	–
R85	220	110	–	–
*R86	225	110	–	–

MAGIC CHEF MODEL NUMBER	AC HEATING ELEMENT		DC HEATING ELEMENT	
	WATTS +/– 10%	VOLTS	WATTS +/– 10%	VOLTS
R86	220	110	–	–
R-87	225	110	–	–
R-88	295	110	–	–
R4412	115/135	110/12	–	–
R4512	135	110	–	–
R4612	115/135	110/12	–	–
R4712	115/135	110/12	–	–
R4812	115/135	110/12	–	–

MAGIC CHEF MODEL NUMBER	AC HEATING ELEMENT		DC HEATING ELEMENT	
	WATTS +/– 10%	VOLTS	WATTS +/– 10%	VOLTS
R5612	115/135	110/12	–	–
R5712	115/135	110/12	–	–
R5812	115/135	110/12	–	–
R6512	115/135	110/12	–	–
R6612	115/135	110/12	–	–
R6712	115/135	110/12	–	–
R6812	115/135	110/12	–	–

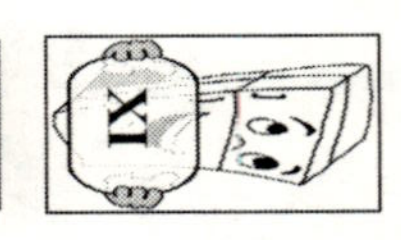

COLD STAR MODEL NUMBER	AC HEATING ELEMENT		DC HEATING ELEMENT	
	WATTS +/– 10%	VOLTS	WATTS +/– 10%	VOLTS
CS3	170	110	–	–
	290	110/12	–	–
	170	110/12	–	–
	250	120	–	–
	250	120/12	–	–
	190	110/12	–	–
CS4	170	110	–	–

COLD STAR MODEL NUMBER	AC HEATING ELEMENT		DC HEATING ELEMENT	
	WATTS +/– 10%	VOLTS	WATTS +/– 10%	VOLTS
	290	110/12	–	–
	170	110/12	–	–
	250	120	–	–
	250	120/12	–	–
	190	110/12	–	–
CS4-I	170	110/12	–	–
	170	110	–	–

COLD STAR MODEL NUMBER	AC HEATING ELEMENT		DC HEATING ELEMENT	
	WATTS +/– 10%	VOLTS	WATTS +/– 10%	VOLTS
(90° bend)	160	110/12	–	–
CS4-I (90° bend)			175	12
CS6	190	110/12	–	–
CS6-I (straight)	250	120	–	–
CS6-I (90° bend)	220	110	220	12
CS8	250	110	–	–
CS8-I	250	110	–	–

COLD STAR MODEL NUMBER	AC HEATING ELEMENT		DC HEATING ELEMENT	
	WATTS +/– 10%	VOLTS	WATTS +/– 10%	VOLTS
CS8-I (slight bend)	220	110	–	–
CS8-I (90° bend)	250	110	–	–
CS10-I (straight)	250	110	–	–
CS10-I (slight bend)	220	110	–	–
CS10-I (90° bend)	260	110	240	12–
(45° bend)	–	–	230	12
	250	120	–	–

COLD STAR MODEL NUMBER	AC HEATING ELEMENT		DC HEATING ELEMENT	
	WATTS +/– 10%	VOLTS	WATTS +/– 10%	VOLTS
(slight bend)	250	110	–	–
(90° bend)	260	110/120	–	–
GL4	170	110	170	12
GL6	250	110	–	–
SL4	170	110/12	–	–
	190	110/12	–	–
SL6	250	120	–	–
	250	120/12	–	–

SIBIR MODEL NUMBER	AC HEATING ELEMENT		DC HEATING ELEMENT	
	WATTS +/– 10%	VOLTS	WATTS +/– 10%	VOLTS
E0821/2A1	140/50	120		
E0821/3A1	140/50	120		
E1221/2A	160/65	240		
E1521/2W and 3W	160/65	117		
	160/65	240		
	160/65	220		
			130	12
E1521/3A1	160/65	220		

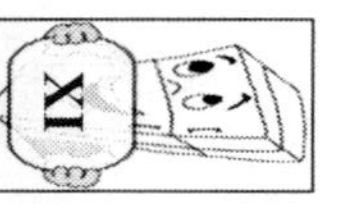
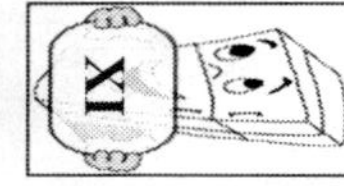

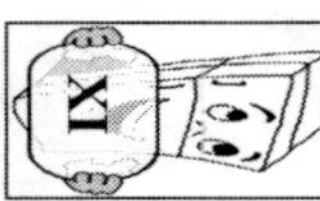

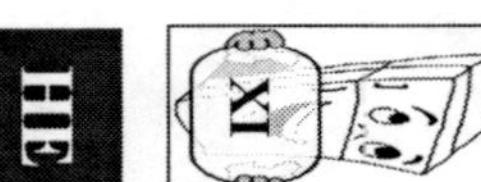

SIBIR MODEL NUMBER	AC HEATING ELEMENT		DC HEATING ELEMENT	
	WATTS +/– 10%	VOLTS	WATTS +/– 10%	VOLTS
E2321/2W	220/95	117		
	220/95	240		
	220/95	220		
E2321/3W	330/95	117		
	220/95	240		
			130	12
			130	24
E2321/2A	330/95	117		

SIBIR MODEL NUMBER	AC HEATING ELEMENT		DC HEATING ELEMENT	
	WATTS +/– 10%	VOLTS	WATTS +/– 10%	VOLTS
	220/95	240		
	220/95	220		
E2321/2A1	220/95	120		
	220/95	220		
E2321/3A	330/95	117		
	220/95	240		
	220/95	220		
E2321/3A1	220/95	120		

SIBIR MODEL NUMBER	AC HEATING ELEMENT		DC HEATING ELEMENT	
	WATTS +/– 10%	VOLTS	WATTS +/– 10%	VOLTS
	220/95	220		
E2321/3A1			130	12
E2322/2W	220/95	117		
E2521/2A1	220/95	120		
	220/95	220		
E2521/3A	220/95	120		
E2521/3A1	220/95	120		
	220/95	220		

SIBIR MODEL NUMBER	AC HEATING ELEMENT WATTS +/– 10%	AC HEATING ELEMENT VOLTS	DC HEATING ELEMENT WATTS +/– 10%	DC HEATING ELEMENT VOLTS
			130	12
E2721/2A	220/95	120		
	220/95	240		
	220/95	220		
E2721/3A	220/95	120		
	220/95	220		
	220	240		
			130	12

SIBIR MODEL NUMBER	AC HEATING ELEMENT		DC HEATING ELEMENT	
	WATTS +/– 10%	VOLTS	WATTS +/– 10%	VOLTS
E2521/3A1	220/95	120		
	220/95	220		
			130	12
E2721/2A	220/95	120		
	220/95	220		
	220/95	240		
	220/95	120		
	220/95	220		

SIBIR MODEL NUMBER	AC HEATING ELEMENT WATTS +/– 10%	VOLTS	DC HEATING ELEMENT WATTS +/– 10%	VOLTS
E2721/3A	200	240		
			130	12
ET150/1	160/65	117		
	160/65	240		
	160/65	220		
ET150/2	160/65	117		
			130	12
	160/65	240		

SIBIR MODEL NUMBER	AC HEATING ELEMENT		DC HEATING ELEMENT	
	WATTS +/– 10%	VOLTS	WATTS +/– 10%	VOLTS
	160/65	220		
ET230/1	220/65	117		
	220/95	220		
	220	240		
ET230/2	220/95	117		
	200	240		
			130	12
RV103	140/50	120		
RV106	160/65	220		

SIBIR MODEL NUMBER	AC HEATING ELEMENT		DC HEATING ELEMENT	
	WATTS +/– 10%	VOLTS	WATTS +/– 10%	VOLTS
RV108	220/95	117		
	200	240		
	220/95	120		
			130	12
RV110	220/95	120		
	200/95	220		
			130	12
			130	24
RV240	165	120		

SIBIR MODEL NUMBER	AC HEATING ELEMENT		DC HEATING ELEMENT	
	WATTS +/– 10%	VOLTS	WATTS +/– 10%	VOLTS
			100	12
RV270			130	12
*RV270	220	120		
RV280			130	12
**RV280	220	120		
S2721/2W1	220/95	120		
	200/95	220/240		
S2723/2W1	220/95	120		
	220/95	220/240		

SERVEL MODEL NUMBER	AC HEATING ELEMENT		DC HEATING ELEMENT	
	WATTS +/– 10%	VOLTS	WATTS +/– 10%	VOLTS
S520	170	100	150	12
S520.013	170	100	–	–
S520.014	185	100	–	–
S530	170	100	150	12
S530.004	170	100	–	–
S530005-010	300	100	–	–
S530.011	170	100	–	–
S530.012-013	170	100	150	12
S619	300	100	–	–

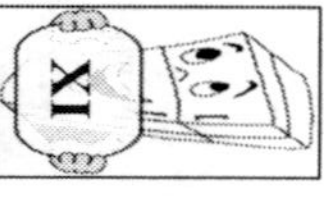

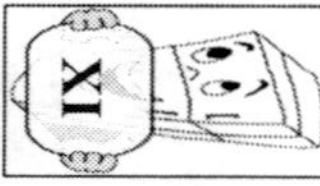

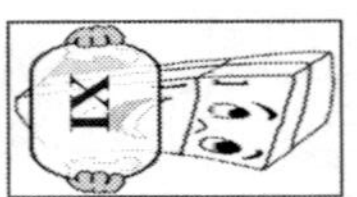
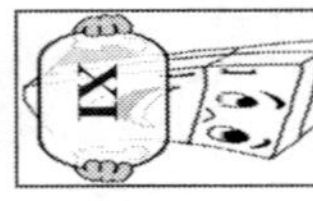

SERVEL MODEL NUMBER	AC HEATING ELEMENT		DC HEATING ELEMENT	
	WATTS +/– 10%	VOLTS	WATTS +/– 10%	VOLTS
S619.007	300	110	–	–
S620	300	110	–	–
S620.013	300	110	–	–
S620.014-015	320	110	–	–
S630	300	110	–	–
S630.013	300	110	225	12
S630.014-015	325	110	215	12
S819	300	110	–	–
S819.007	300	110	–	–

SERVEL MODEL NUMBER	AC HEATING ELEMENT		DC HEATING ELEMENT	
	WATTS +/– 10%	VOLTS	WATTS +/– 10%	VOLTS
S820	300	110	225	12
S820.001-13	300	110	–	–
S820.014	325	110	215	12
S820.015	–	–	215	12
S830	300	110	–	–
	325	110	–	–
S830.007	300	110	225	12
S830.008-013	300	110	225	12
S830.014	185	110	215	12

SERVEL MODEL NUMBER	AC HEATING ELEMENT		DC HEATING ELEMENT	
	WATTS +/– 10%	VOLTS	WATTS +/– 10%	VOLTS
S1521	185	110	–	–
S1531	185	110	175	12
S1621	325	110	–	–
S1631	325	110	215	12
S1831	325	110	215	12

Remember, an RVRN member offers other services for your RV.

YOUR NOTES

TRAVELER MODEL NUMBER	AC HEATING ELEMENT WATTS +/– 10%	AC HEATING ELEMENT VOLTS	DC HEATING ELEMENT WATTS +/– 10%	DC HEATING ELEMENT VOLTS
100240	300	117	–	–
120240	300	117	–	–
20190	110	110/12	–	–
20290	110	110/12	–	–
30140	140	110	–	–
	140	110	–	–
	140	110/12	–	–
	140	110/12	–	–

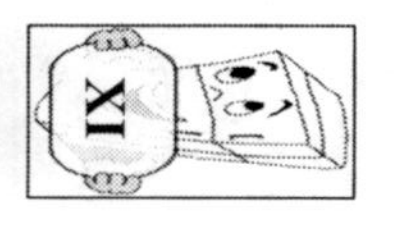

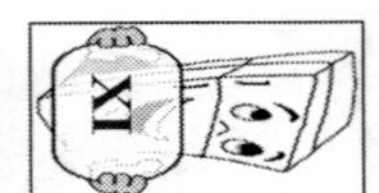

TRAVELER MODEL NUMBER	AC HEATING ELEMENT WATTS +/– 10%	AC HEATING ELEMENT VOLTS	DC HEATING ELEMENT WATTS +/– 10%	DC HEATING ELEMENT VOLTS
	140	110	–	–
30190	140	110	–	–
	140	110/12	–	–
	140	110/12	–	–
	140	110	–	–
30240	140	110	–	–
	140	110/12	–	–
30290	140	110	–	–

TRAVELER MODEL NUMBER	AC HEATING ELEMENT		DC HEATING ELEMENT	
	WATTS +/– 10%	VOLTS	WATTS +/– 10%	VOLTS
	140	110/12	–	–
40140	170	110	–	–
	170	117	–	–
	170	110	–	–
	170	110/12	–	–
	170	110/12	–	–
	170	110/12	–	–
40190	170	110/12	–	–

TRAVELER MODEL NUMBER	AC HEATING ELEMENT		DC HEATING ELEMENT	
	WATTS +/– 10%	VOLTS	WATTS +/– 10%	VOLTS
	170	117	–	–
	170	110	–	–
	170	110	–	–
	170	110/12	–	–
	170	110/12	–	–
60140	200	110	–	–
	200	117	–	–
60190	200	117	–	–

TRAVELER MODEL NUMBER	AC HEATING ELEMENT		DC HEATING ELEMENT	
	WATTS +/– 10%	VOLTS	WATTS +/– 10%	VOLTS
	200	110	–	–
60240	200	117/12	–	–
	200	117	–	–
60290	200	117/12	–	–
70140	280	110	–	–
70240	280	110	–	–
80140	340	110	–	–
80240	300	117	–	–

SECTION XII

RV REFRIGERATOR PARTS GENERAL INFORMATION

1. Orifice—Sometimes the orifice gets dirty. Soak the orifice in an alcohol-based solvent for a few minutes, then allow air to dry. The size of the orifice is also an important factor (see figure 12-1). Do not put any type of foreign object into the hole of the orifice.

Figure 12-1.
Orifice (Norcold)

Below is a chart showing some of the Dometic orifices.

MODEL	ORIFICE SIZE
RM2607	no. 53
RM2611	no. 53
RM2807	no. 58
RM2811	no. 58
RM3607	no. 53
RM3807	no. 58
RM3662	no. 58
RM3663	no. 58
RM3862	no. 58
RM3863	no. 58
RM4872	no. 58
RM4873	no. 58
S1521	no. 43
S1531	no. 43
S1621	no. 58
S1631	no. 58
S1821	no. 58
S1831	no. 58

2. Thermocouple—should generate 25-35 millivolts (see figure 12-2). Chcck with a meter that reads millivolts. If it does not, replace thermocouple.

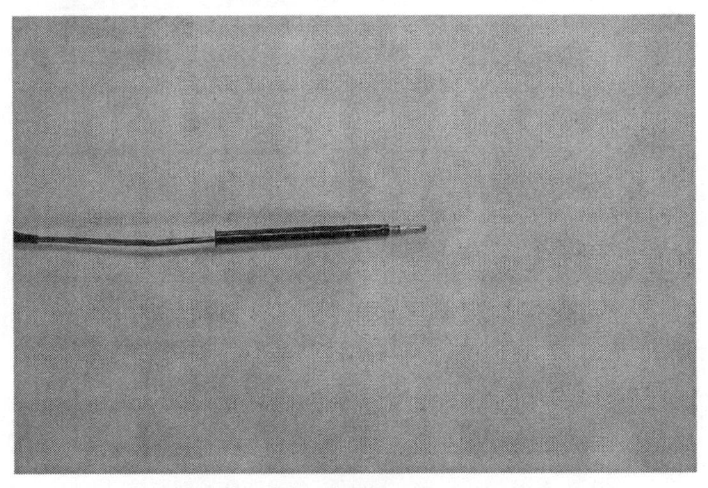

Figure 12-2. Thermocouple (found above burner)

QUOTE

"The RVRN has a proven track record of teaching safe and sound practices of reconditioning, and that is why we have been involved with the organization and recommend it highly because of the integrity of the organization. I have never met any other people who are as knowledgeable or who care as much about safety and protecting the consumer through proper education as Roger and Onna Lee Ford do."

—Mike and Cindy T., RVRN members from Florida, 2009

3. Burner—The slots in the burner must be located right below the flue tube (see figure 12-3). It needs to be cleaned at least once a year.

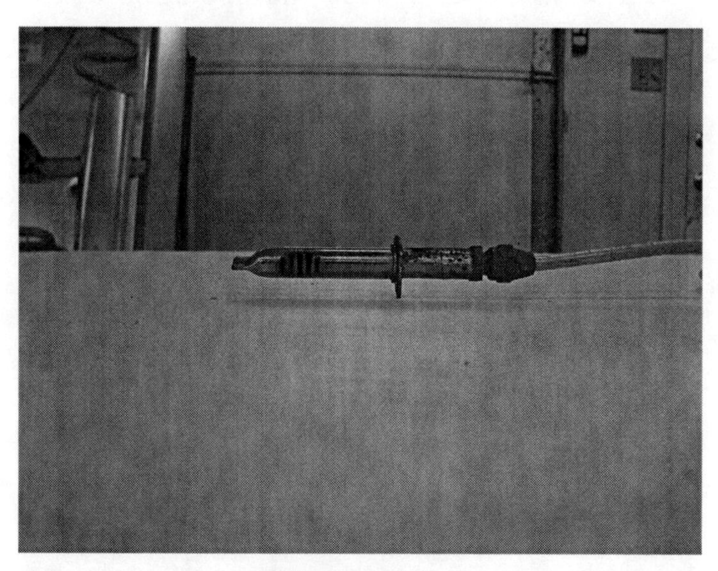

Figure 12-3.
Burner (Norcold)

QUOTE

"Businesses and consumers are increasingly interested in 'going green.' They are seeking ways to decrease their carbon footprints and save money."

—*HVACR Business* magazine, January 2009, 14

4. Flue Baffle—The flue baffle, which is located inside the flue, requires cleaning at least once a year, depending on gas usage in that year (see figure 12-4). The position of the baffle is critical to the operation of the refrigerator.

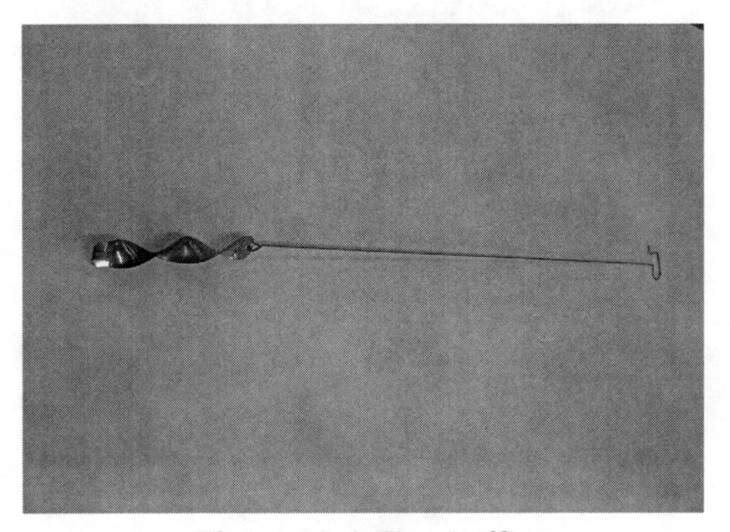

Figure 12-4. Flue baffle

QUOTE

Jason, a master certified RV technician from Nevada, said, "I was amazed at how much I did not know about RV refrigerators. Everything I thought I knew about the operation of the cooling unit was wrong, and everybody I had talked to before I took Ford RV's training knows all the same wrong information."

Below is a chart indicating the distance the Dometic baffle should be from the top of the burner.

DOMETIC MODELS	BAFFLE DISTANCE
RM2607	1 5/8"
RM2611	1 5/8"
RM2807	1 3/4"
RM2811	1 3/4"
RM3607	1 5/8"
RM3807	1 3/4"
RM3662	1 7/8"
RM3663	1 7/8"
RM3862	1 7/8"
RM3863	1 7/8"
RM4872	1 7/8"
RM4873	1 7/8"
S1521	2 1/4"
S1531	2 1/4"
S1621	1 5/8"
S1631	1 5/8"
S1821	1 5/8"
S1831	1 5/8"

5. Thermistor—To check the correct function of the thermistor, disconnect it from the 2 in. terminal on the lower circuit board (see figure 12-5). Submerse the thermistor for 2-3 minutes in a glass of ice water. The reading should be anywhere from 7,000-10,000 ohms if it is functioning properly. If not, replace thermistor. You will need a multimeter to read ohms.

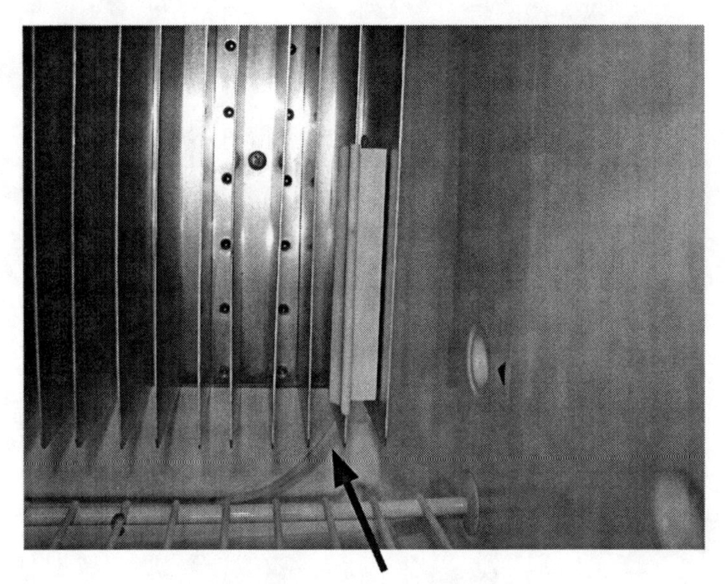

Figure 12-5. Thermistor

QUOTE

"A happy customer is our best investment towards advertising."

—Onna Lee Ford

SECTION XIII
BELIEVE IT OR NOT

Many owners have, on the advice of others uneducated in this field, tried the remedies. When these remedies failed to produce the desired results, the units were brought to Ford RV Refrigeration for proper diagnosis and repair.

Remember, it's easy when you know how. It's all about education.

The following are "remedies" reported to Ford RV Refrigeration.

MYTH	TRUTH
Chemical has separated.	When unit has remained idle for a long time, chemicals will separate a little. However, once heat is applied to boiler, chemicals remix and normal operation will resume.
Chemical has jelled.	There are no chemicals in unit that can jell.
Chemical has crystallized.	Chemicals cannot crystallize as long as unit is fully charged and sealed.
Air pocket has formed in unit.	Air cannot enter unit unless a hole is present, in which case all pressure inside unit would be released. The problem would then be a leak, not an air pocket.

Note: No offense is intended to anyone in the areas indicated that follow. We use these areas because the information came from someone that resides there. This is only to be viewed with humor although these remedies were actually reported to us by individuals that tried them.

There is no logical reason for any of the following remedies to work. If any ever did work, in all probability, the cooling unit was never bad. Either dirt was in the gas line or orifice, or there was a bad connection in the wiring or one of the controls. Throwing, dropping, torching, and hammering on the unit could be very dangerous unless no charge exists in the unit, in which case a leak is present, and none of these "remedies" will repair a leak.

If you have an "old-time remedy," please send it to Ford RV Refrigeration, 1746 Big Bear Hwy., Benton, KY 42025. Please include you location and how you would like your name listed in the event it is used in any of our other publications.

By sending this information, you

1. are giving permission for its use in any manner,
2. and understand there will be no payment to or from you for the use of this information.

KENTUCKY CONFUSION CONTUSION

**Driving as fast as you can,
along a bumpy road,
with refrigerator in the bed
of a pickup truck**

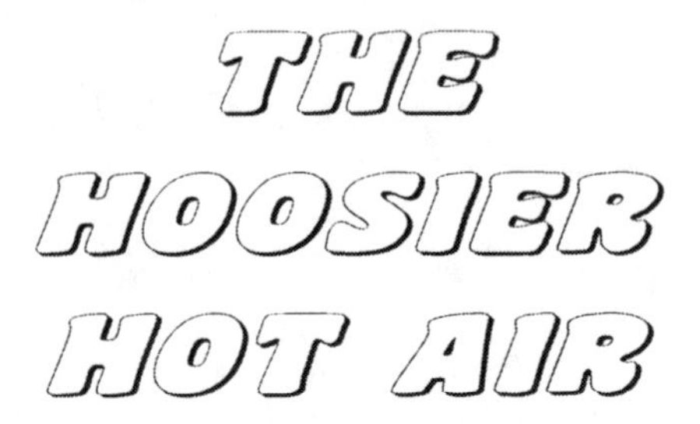

Turning unit upside down for twenty-four hours

THE RAZOR-BACKED ROCKER

Rocking refrigerator for tweny-four hours

THE ROCKY MOUNTAIN ROCK-N-ROLLER

Rotating refrigerator from side to side

THE SHOW-ME SHOCKER

Applying electrical shock to the refrigerator

THE BATTIN' BUCKEYE

Grand-slamming all sides of the refrigerator

THE BAYO BASH

Hitting all the coils on the unit repeatedly

THE FLORIDA FLING

Throwing the cooling unit

THE BUFFALO DROPPINGS

Dropping the cooling unit repeatedly

www.rvrefrigeration.com

WHEN YOU'VE REACHED THE END OF YOUR ROPE

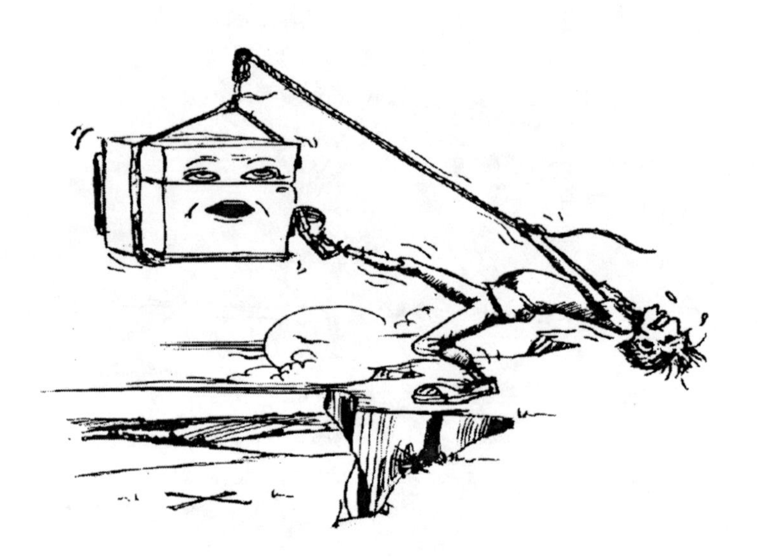

YOU NEED

An RVRN refrigerator specialist

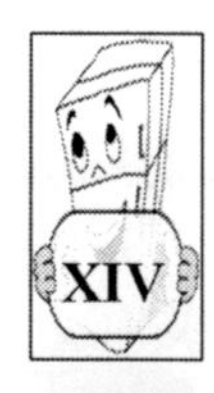

SECTION XIV
GLOSSARY

absorption refrigeration	Refrigeration that creates temperatures by using the cooling effect of a refrigerant being absorbed by a chemical substance.
alternating current (AC)	Electrical current in which direction of flow alternates or reverses. In the case of 60-cycle (hertz) current, for example, direction of flow reverses every 1/120th of a second.
ambient	Temperature of surrounding air or fluid.
boiler	Closed container in which a liquid may be heated and vaporized.
charged	Cooling unit has the correct formula needed to operate properly.
condense	The process of changing a gas into a liquid.
condenser	An apparatus for liquefying gases or vapors.
cooling unit (core)	Metal piping on the back of the refrigerator that houses the chemicals used to create the cooling system.

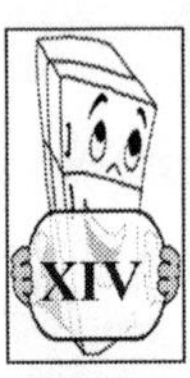

Dalton's law	Vapor pressure created in a container by a mixture of gases is equal to the sum of the individual vapor pressures of the gases contained in the mixture.
direct current (DC)	An electric current flowing in one direction.
gas	Vapor phase or state of a substance.
heat element	A heat-producing electrical device, which operates off 12-volt DC or 110-volt AC.
level refrigerator	Placing a level in the refrigerator from side to side and front to back.
liquid	Substance whose molecules move freely but do not tend to separate indefinitely as do those of a gas.
LP fuel	Liquefied petroleum. Used as a fuel gas.
odor	That property of air contaminants that affect the sense of smell.
ohmmeter	Instrument for measuring resistance in ohms.
Ohm's law	Mathematical relationship between voltage, current, and resistance in an electric circuit. The law is stated as follows: voltage (e) equals amperes (i) times ohms (r), or $e = i \times r$.

orifice	Accurate-sized opening for controlling flow.
plug	A condition caused in the boiler tube when the unit is run unlevel.
pressure	Energy impact on a unit area; force or thrust on a surface.
propane	Volatile hydrocarbon used as a fuel.
separator	Device to separate one substance from another.
temperature	Degree of hotness or coldness as measured by a thermometer.
thermometer	Device for measuring temperatures.
watt	Unit of electrical power.

QUOTE

"The quality of training is very good.
The Ford RV crew, without exception,
I found to be courteous, knowledgeable,
and professional at all times."

—BJ, an RV service technician for thirty years from Utah, 2008

If you have determined that the cooling unit needs to be reconditioned, contact us to find the nearest RVRN member who can take care of this for you. In doing so, you will save money, receive the industry's best warranty, plus you help save our environment by keeping your cooling unit and refrigerator out of the landfills!

Remember,
it's all about education.
It's easy when you know how!

RV owners, receive a
hands-on basic training
on your RV refrigerator.

Contact our office for more information.

If you know anyone who would be interested in learning the procedures for reconditioning the cooling unit on RV refrigerators, please have them contact

Ford RV Refrigeration
Training Center
1746 Big Bear Hwy.
Benton, KY 42025
270-354-9239
fordrv@rvrefrigeration.com
www.rvrefrigeration.com

SECTION XV
REFRIGERATOR RECALLS

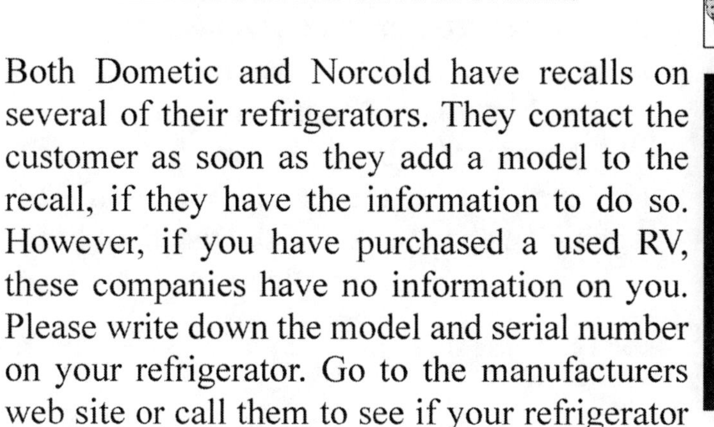

Both Dometic and Norcold have recalls on several of their refrigerators. They contact the customer as soon as they add a model to the recall, if they have the information to do so. However, if you have purchased a used RV, these companies have no information on you. Please write down the model and serial number on your refrigerator. Go to the manufacturers web site or call them to see if your refrigerator is under recall. If it is not, be sure to check back periodically as they add units occasionally. If it is they will instruct you as to what you should do next.

If your unit does fall under the
recall,
SHUT THE REFIGERATOR
COMPLETELY OFF!

Note: If you purchased your RV used, it may or may not have the kit on it. Go to the service door (outside of RV) that is behind the refrigerator. Open the service door. You are now viewing the back side of the refrigerator and can see the wiring, the cooling unit, etc.

FOR DOMETIC REFRIGERATORS

If you do not see a metal cover (heat shield), with the screws on the side and bottom (screws must be there) over the burner, the kit has most likely not been installed or was not installed properly. However, to be extra sure, have a certified technician check to see if the recall kit has been installed.

FOR NORCOLD REFRIGERATORS

If the recall kit *has not* been installed, the insulation canister (boiler pack) will *not* have a collar installed at the bottom by the switch hole. If the collar is installed at the bottom, it would appear that the recall kit *has been* installed. Again, to be extra sure, have a certified technician double–check it.

If the kit *has not* been installed, or you are not sure, do the following:

SHUT THE REFRIGERATOR COMPLETELY OFF!

1. You will need the following information:
    a. For Dometic refrigerators:
        i. Model number, serial number, VIN, RV manufacturer
    b. For Norcold refrigerators:
        i. Model number on refrigerator
        ii. Serial number on refrigerator
        iii. Serial number on the back of the cooling unit
        iv. VIN of the RV
        v. Date RV was manufactured
        vi. Date you purchased your RV

2. Contact Ford RV Refrigeration or an RVRN refrigeration specialist in your area or your local RV technician.

3. Schedule a time to have the recall kit installed on your refrigerator.

4. Once you are sure the recall kit has been installed, you may resume operation of the refrigerator.

There should be no charge to you for the recall kit or the installation. All paperwork and payments are handled between the technicians and the manufacturers. If your RV is mobile, there may be a charge for a service call.

RV PRO
Aftermarket Products and Profits
The complete information source for RV Aftermarket Professionals!
Are you the RV Products Buyer or Service Tech for your company?
If so, YOU should be reading RV PRO every month!
Don't miss an issue!
Order your FREE subscription TODAY!
Look forward to...
• Vivid pictorials of the latest products
• Feature articles showcasing the most up-to-date industry trends
• Features profiling outstanding dealerships and aftermarket retailers
• Dealer Trends: What dealers are saying about industry developments
• Innovative sales, marketing & merchandising solutions
• Advice on supercharging your business operations to help you fatten your bottom line
...every issue in RV PRO!
To subscribe visit us online at
www.rv-pro.com
Or call us toll-free: 1-800-870-0904
Use Promotion Code ERVREF

WHY SHOULD I LOOK FOR A CERTIFIED TECHNICIAN?

A certified technician is

- educated,
- committed,
- qualified
- experienced
- interested in the benefit of the customer,
- honest, and stands behind his/her work.

They were glad they did!

Jerry and Barb from Illinois
Irene from Kentucky

INDEX

INDEX

INDEX

INDEX

INDEX

INDEX

INDEX

INDEX

INDEX

QUOTE

"Roger has a great storehouse of knowledge.
Or it could be said that he has been
around the block *many* times."

—Sid, a certified specialist in commercial
and industrial ammonia refrigeration
in Tennessee, 2007

QUOTES

"I would like to thank you both once again for sharing your kindness and knowledge with me while Jane and I were there last week. I was only looking to get a little bit of hands-on training on RV refrigeration, but I came away with a whole lot more than I could have expected. Your kindness, inspiration, and enthusiasm were more than I had bargained for, and for that, many, many thanks. Together with your personalities and your strong commitment to your business ethics, I would wholeheartedly recommend Ford RV Refrigeration to anyone."

—Jim, RV owner from Maryland who received a five-hour training on controls from Ford RV

"I have really enjoyed this course as I learned a great deal from very knowledgeable people who were truly interested in my development and not in their own pocketbook."

—Wade, executive in a nuclear plant from Idaho, 2007

QUOTE

About Ford RV Refrigeration Training Center, "A very welcome atmosphere, positive attitudes, clean environment, very good and thorough training, all explanations very thorough. I would recommend this school to anyone wishing to learn this technique regardless of age or occupation. I have over fifty years in HVACR experience. The Ford Procedures Manual is your road map to success."

—Bob, RVRN member from Tennessee, 2009

DO YOU NEED RV OR MARINE
PARTS AND ACCESSORIES?

Go to our Web site:
www.rvrefrigeration.com/catalog.asp
and order just about anything you can imagine to make your weekends, vacations, travel, etc., more enjoyable.

Shop, place your order and pay online, indicate where you would like it to be shipped, and within days you will have your order.

Clothesline **Water Toys** **Sea Clamp** **Multi-max Adapter**

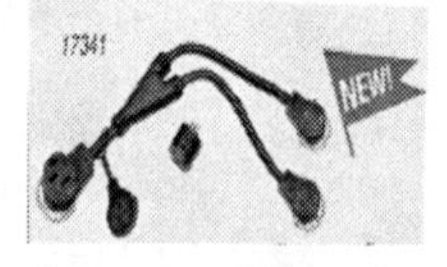

and much, much more!

QUOTES

Sylvia from Illinois said, “I have had a meaningful experience. I have learned a great deal from everyone involved in my training. I came into this, knowing virtually nothing relating to this procedure. I am leaving feeling confident that I can take what I have learned back home and being working on units.”

Sylvia was our first woman technician. Her background is in office/management, manufacturing, CDL, and as an operating engineer.

Kevin from North Dakota said, “I found this to be very rewarding to me. Everyone has made me feel comfortable being here. Everyone has explained everything probably better than needed to be, but am grateful for that too, it helps. I thank you very much.”

Kevin has been in the RV service business for fourteen years. He is also certified in high pressure steam boilers.

QUESTIONS?

Please use this page to write down anything you question or can't find an answer to in this manual.

Send to:
fordrv@rvrefrigeration.com
or
1746 Big Bear Hwy.
Benton, KY 42025

In the future, we will use these on a question-and-answer page at our Web site and in future manuals. By sending these, you automatically agree to the same conditions found on page 200.

Questions Continued

GO GREEN-GO RNRN

Training Application

(PLEASE PRINT WITH BLUE INK)
Or fill out online @
www.rvrefrigeration/training.asp

NAME______________________________

ADDRESS___________________________

CITY___________________STATE_______

ZIP____________________PHONE_______

E-MAIL____________________________

BIRTH DATE________________________

The following questions are only to give us an idea of your background. They are no requirements to obtain the training.

1. How many years have you been in the RV business? ____________________

2. How many years have you worked on RV refrigerators? ____________________

3. If not in the RV business, what field of business are you in? ____________________
   How long? ____________________

4. Do you have any certifications?
   ______Yes ______No

   If so, what are they? ____________________
   ____

TRAINING APPLICATION

5. Do you have any experience with a multimeter?

 _____Yes _____ No

6. Do you have welding experience?

 _____Yes _____ No

7. Do you plan to join the RVRN Association?

 _____Yes _____ No

Please tell us your reasons for

1. Wanting to join the RV refrigerator reconditioning industry.
2. What your goal is upon completion of training.

Minimum—100 words
Maximum—200 words

____

____

____

____

____

_________________________ _______________

Signature Date

Add a sheet if necessary.
Mail to:
Ford RV Refrigeration
1746 Big Bear Hwy.
Benton, KY 42025

GO GREEN-GO RNRN

This guide was written at the request of many RV owners. It is a tool to help you save money on repairs and unnecessary replacements. It contains step-by-step instructions, charts, pictures, places for your own notes, myths and remedies, and much more. You can even find information on the training we offer should you wish to open your own business or add to an already existing one.

How do the Ford Procedures compare to those who mass-produce new and reconditioned RV refrigerator cooling units? The Ford Procedures are successful because we check every step of every procedure for every repair rather than only spot-checking.

If you need refrigerator service or want to be successful at reconditioning, part-time or full-time, make great profits, save the consumer money, and provide the best warranty in the RV industry.

GO GREEN WITH US: Call 270-354-9239.
Ask about our lifetime warranty!

LaVergne, TN USA
23 March 2010
176931LV00004B/143/P